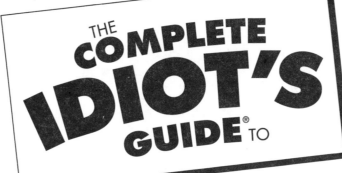

THE COMPLETE IDIOT'S GUIDE® TO

Remodeling Your Bathroom Illustrated

by Dan Ramsey

ALPHA

A member of Penguin Group (USA) Inc.

For Jude, my soul mate.

Publisher: *Marie Butler-Knight*
Product Manager: *Phil Kitchel*
Senior Managing Editor: *Jennifer Chisholm*
Senior Acquisitions Editor: *Mike Sanders*
Development Editor: *Lynn Northrup*
Production Editor: *Megan Douglass*
Copy Editor: *Ross Patty*
Illustrator: *Chris Eliopoulos*
Cover/Book Designer: *Trina Wurst*
Indexer: *Angie Bess*
Layout/Proofreading: *Becky Harmon, Donna Martin*

Contents at a Glance

Appendixes

Contents

Appendixes

Foreword

Remodeling is a dirty job. It's typically not clean work like building a home. More often it's dirty work like tearing out 50-year-old walls or climbing around under a house to change the plumbing. As a remodeling contractor, I don't relish the time spent getting dirty. In fact, I prefer most anything to getting dirty. And my fellow remodeling contractors typically feel the same.

That's why we all attempt to work smart. We think the project through, plan the job out, and make sure we have all the needed materials and tools nearby so we can get in and get out. The smarter our planning the less we have to get dirty.

If you're thinking about remodeling your home's bathrooms, you'd be wise to follow our example: Plan more and get dirty less. It's good advice.

My friend Dan Ramsey dramatically cuts the dirty work of remodeling bathrooms with this book, *The Complete Idiot's Guide to Remodeling Your Bathroom Illustrated*. He not only tells you how to get the job done with the least effort, he also shows you with hundreds of photographs and drawings. You'll see what you're getting into.

You'll also know about the pitfalls of remodeling the throne room. He includes dozens of practical tips that he, I, and many other contractors follow to tackle the job. You'll also learn how to get permits, financing, and the other behind-the-scenes jobs done that can be more dreadful than crawling around under an old house.

So before you start pulling out faucets and cabinets, read *The Complete Idiot's Guide to Remodeling Your Bathroom Illustrated* from cover to cover. Don't be fooled by the conversational tone. There's real information here, information that can help you not only with the dirty work but also show you how to get the job done for the least time and money. And if there's a technical term you don't quite get, turn to the glossary at the back for a clear explanation and maybe a photo. There are additional resources in another appendix, too. Dan's book tells you everything you need to know to get a dirty job done with the least effort and the greatest results!

Enjoy your remodeled bathroom!

Greg Dunbar
Greg Dunbar Construction

Introduction

Last year, we homeowners spent more than *120 billion dollars* on remodeling! Many of our projects included the bathroom—facelifts, makeovers, and additions. Why? We know that remodeling a bathroom can pay off in increased value and functionality. That's why so many of us are taking advantage of lower interest rates and remodeling the bath. You can, too.

The Complete Idiot's Guide to Remodeling Your Bathroom Illustrated shows you how to design, plan, fund, and remodel all sizes and types of bathrooms without hiring a contractor. Profusely illustrated with hundreds of photos and drawings, this book *shows* you each step in the bathroom planning and remodeling process, from scratching out ideas to adding decorative items to your finished bathroom. Each of the projects includes clear step-by-step instructions and illustrations. Along the way you'll also get proven tips from remodeling pros to save you countless hours and dollars. And you'll be proud of the results!

Who am I to tell you how to remodel your bathroom? Besides being a licensed remodeling contractor, I'm a longtime homeowner who has done many of my own bathroom remodeling jobs in various homes. I'm also the author of *The Complete Idiot's Guide to Building Your Own Home* and *The Complete Idiot's Guide to Finishing Your Basement Illustrated*. I'll share proven tips I've learned—sometimes the hard way—and add some from other remodelers and suppliers so you get the most from every dollar and every hour you spend beautifying your home's bathrooms. Included are photos of *real* bathroom remodeling jobs and finished projects. You'll see what you're getting into.

Even if you've never swung a hammer or tightened a faucet fitting you'll be comfortable remodeling your bathroom as you read *and see* this book. Save thousands of dollars and enjoy the pride of saying "I did it myself!" Remodel your bathroom.

How to Use This Book

This book is presented in a very logical structure to make it easy to find things when you need them. Let's take a closer look at each part.

Part 1, "Planning Your New Bathroom," helps you do just that. The more time you spend planning the less time you'll spend on remodeling. Contractors know this. This first part shows you the differences between a bathroom facelift, a makeover, and an addition as well as how to plan and design the type of bathroom remodel that works for you and your budget.

Part 2, "Remodeling Your New Bathroom," focuses on the three primary types of bathroom remodels—facelift, makeover, and addition—with specific guidelines and step-by-step instructions. For those tackling simpler remodels or who have some remodeling experience, this part gives you everything you need to get the job done.

Part 3, "Remodeling Bathroom Systems," digs deeper into how each of the primary bathroom systems—plumbing, wiring and lighting, heating and ventilation, walls and flooring, cabinets, fixtures, and accessories—work. Included are extensive instructions and illustrations to walk you through all types of bathroom system projects.

In addition, you'll find an illustrated Glossary at the back of the book that defines nearly 200 terms, and a list of valuable resources for further study.

Extras

Along the way, practical sidebars show you the safe and smart way to do things, define words and terms you may not be familiar with, point out any dangers or pitfalls, and give you other bits of helpful information. *The Complete Idiot's Guide to Remodeling Your Bathroom Illustrated* makes the job easier—and maybe even fun!

Potty Training
Here are some valuable tips from remodeling contractors on how to do the job right the first time.

Waste Lines
Don't throw money down the drain. Take good advice on how to save time and money on bathroom remodeling.

Bathroom Words
What does *that* mean? Here you'll find a concise definition of important bathroom remodeling terms in context. Also check the illustrated Glossary for more definitions.

Heads Up!
Bathrooms can be dangerous places to work! Here's how to stay safe as you work with structural, plumbing, electrical, and other systems.

Acknowledgments

Remodeling is not only a major industry, it is a valuable alliance. I have learned many priceless lessons from my fellow do-it-yourselfers and remodelers that I'll share with you in this book. Let me acknowledge some of them:

Ted Bozzi and Melissa Marciano of National Kitchen & Bath Association for technical and photo assistance.

John Mandel of USG Corporation for valuable illustrations and assistance.

Daniel Reif of homeplanner.com for the opportunity to use QuickPlanner, a great home design tool.

Greg Dunbar of Greg Dunbar Construction who always has a tip and an anecdote on virtually any remodeling topic!

Bill and Fran Schatz for the use of their bathroom. And to Loren and Aaron Luedemann, their able remodelers.

Dave Schrock of Dave's Dwellings, Inc. of Aurora, IL (www.basementideas.com), who supplied useful photographs.

Brian Ellis of ECI Builders of Farmington Hills, MI (www.ecibuilders.com), who supplied photos and tips.

Bill Farmer of Mendo Mill, Willits, CA, and Bob Doty of Mendo Mill, Ukiah, CA, who helped me illustrate this and prior books.

Jack Hulsey of Premier Design & Construction Inc. of Ukiah, CA, who gave me access to homes under construction.

Scott Harris of ART, Inc., developers of Chief Architect, for loaning me their professional design software.

Thanks to the Fix-It Club (www.FixItClub.com) for additional information and support.

Also thanks for technical advice and illustrations from Kohler, Co., Tile Council of America, Stanley Works, Delta Faucet Co., Genova Inc., and the Cooper Group.

Thanks to Mike Sanders, Lynn Northrup, Megan Douglass, Ross Patty, and the Alpha Books team for sharing their quality skills. It's always a pleasure.

Special Thanks to the Technical Reviewer

The Complete Idiot's Guide to Remodeling Your Bathroom Illustrated was reviewed by an expert who double-checked the accuracy of what you'll learn here, to help us ensure that this book gives you everything you need to know about bathroom remodeling. Special thanks are extended to Greg Dunbar, who has more than 20 years' experience building and remodeling homes as a licensed contractor.

Trademarks

All terms mentioned in this book that are known to be or are suspected of being trademarks or service marks have been appropriately capitalized. Alpha Books and Penguin Group (USA) Inc. cannot attest to the accuracy of this information. Use of a term in this book should not be regarded as affecting the validity of any trademark or service mark.

In This Part

Part 1

Planning Your New Bathroom

Is it true that Einstein created his greatest theories in the bathroom? Probably not. But it's always been a popular place for the contemplation of changes …

"Okay! I'll be out in a minute! If you're in such a hurry, go to the neighbor's house!"

"It would sure be nice to have a hot tub and a TV in here."

"I'm *really* getting tired of orange floor tile!"

"I wonder what the Queen's loo looks like."

Then one day it occurs to you to remodel your bathroom. But where to start? Right here! The first section of this book will take you through the planning process to help you decide what you want, how much it will cost, and what you'll have when you're done.

Meantime, it's something fun to read while you're in the john.

In This Chapter

- ◆ Taking the easy road to a better bathroom
- ◆ How the modern bathroom has evolved
- ◆ Options for half, full, and extended baths
- ◆ Things to think about when planning a bathroom facelift

  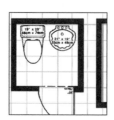  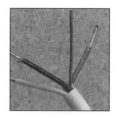 

Chapter 1

Bathroom Facelifts

Has your bathroom seen better days? Are you interested in giving your bathroom a fresh new look, but are not sure how to start? Whoa! Don't start tearing out walls and fixtures just yet. There are dozens of things you can do to make your home's bathroom nicer without removing things. Or maybe just a few things. You can apply a little "Bathroom Botox" in a few hours or over a weekend to take out a few wrinkles here, a tuck there, and make the throne room look young again!

This first chapter is an idea chapter. You'll see how numerous average and below-average bathrooms are converted into model baths without moving major components. Later chapters will show you exactly how to do these things. For now, you want ideas. So sharpen a pencil, grab some scratch paper, and step into the lavatory.

Bathroom Basics

We in modern society are totally spoiled. A special room—or numerous rooms—dedicated to personal hygiene is a luxury previously only known to potentates. In today's housing world, anything less than one bathroom for every two household members seems uncivilized.

On the other hand, one of the reasons that society has advanced is the bathroom. Moving it indoors, engineering it for sanitation, using easier-to-clean materials, and adding amenities has made us more sensitive to personal hygiene. We have fewer diseases and pestilences. We live longer, healthier lives because of indoor plumbing. It's much better than the alternatives and, frankly, we should be thankful. I certainly am!

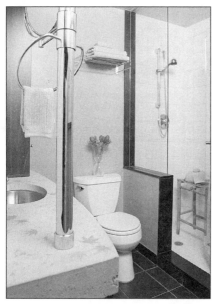

Bathrooms have extensively evolved since the days of public baths.

Evolution of the Throne Room

Hygiene has always been an essential part of society. Though indoor plumbing is a modern development for the average person it has served rulers and the rich for thousands of years. It was just too dang expensive to redirect water flow and to construct sewer systems! Only the rich and famous could afford the costs.

Then some entrepreneur or official up for election decided to share the costs among many, and the public bath was born. It included huge communal bathtubs near a source of flowing water. Individuals, families, and even strangers bathed together—though good manners forbade bathers from staring at each other. In many societies today, communal bathing continues to be a part of daily life.

Cabinets are fully attached to walls in which the plumbing is installed.

As populations grew, private bathing became more important to individuals. Houses often had inside bathing tubs, but the water source was still outdoors, meaning someone had to lug the water in pails, heat it on a stove, and pour it into the tub and jump in before it cooled. Servants became the rage. Of course, everyone still had to go outdoors for other bathroom duties, so *outhouses* or *privys* were still a necessity.

Bathroom Words

An **outhouse** is any habitable structure outside the main building, though typically meaning a structure with a toilet. A **privy** is a private room or other structure including an outhouse with toilet.

Along came the industrial age when cities grew and communal bathing and outhouses were getting out of hand. Fortunately, industry was able to figure out how to make common things and even some uncommon ones at lower cost. More people could afford pipe and new-fangled faucets. Simple by modern standards, the indoor bathroom soon became the rage. Every house had to have one.

Today, ingenuity continues its trek toward more and more bathroom amenities: Vichy showers, bidets, hydrotherapy tubs, and music systems. Hey, why not?! Bathrooms are more than functional, they are therapeutic, a place to go beyond basic personal hygiene toward comfort and relaxation. Chances are, there are many options available to your bathroom that weren't economically feasible when the home you may live in was built. You can upgrade your bathroom(s) with features and amenities that can add value to your life. Go for it!

A Look Behind the Walls

No matter what you do to change the look or function of your home's bathroom there are some basic things you need to know. Coming chapters, especially those in Part 2, will give you the details. For now, let's look inside your home's walls to see what's going on.

First, the bathroom needs fresh water. Actually, there are two fresh water systems in your home: hot and cold. You knew that. Water from a public water system or a private well or spring system is delivered under pressure to your residence in a single pipe. Part of it is diverted to one or more water heaters. It then flows through pipes to fixtures such as faucets and toilets. The pipes that carry the water are installed in the walls, their location depending on the shortest path.

Second, waste water from the sink or tub and from the toilet must be removed from the residence. So bathrooms have waste removal systems as well, located in the walls and in the floor. Eventually, it all winds up in a waste pipe that allows it to dump into a sewer or septic system. Sewers are public utilities with large pipes located under the road in front of your house. Septic systems are private disposal systems, typically a septic tank and drain fields located in your yard.

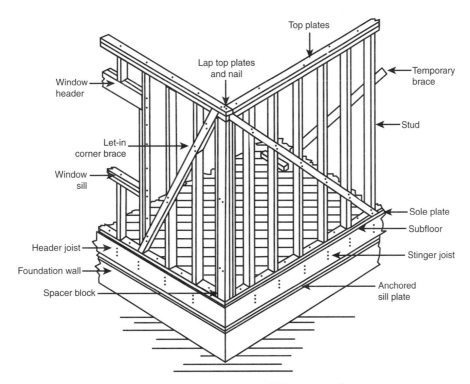

Components of a typical home's wall.

Potty Training

Most homes use common plumbing walls. That is, bathrooms often are located next to or near each other, sharing a common wall that includes pipes for both rooms. It cuts construction costs.

Third, there's a ventilation system that allows gases to escape from the waste plumbing. Together, the waste system is called

drain-waste-vent or DWV. You'll learn much more about these systems in coming chapters.

Also in the walls is the electrical system: wires that bring the electricity to receptacles (outlets or plugs), fixtures (such as lighting), and switches that control the fixtures. Most wiring runs along and through vertical wooden wall supports called studs, typically located 16-inches apart. Above and below the studs are other wooden components called plates. Covering the studs are sheets of plaster called drywall or wooden slats (lath) covered with dried plaster.

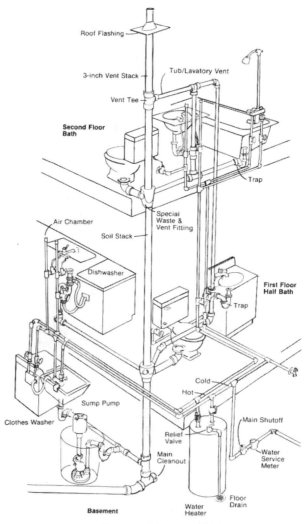

Components of a typical home's plumbing system.

Now that you can envision what's behind the walls of your home's bathrooms you can better understand the limitations and opportunities you may have when remodeling. You know that moving a toilet even a few feet, for example, means you will have to move the cold water line *to* it as well as the waste line *from* it. So maybe it's not so bad where it is.

Use your imagination to picture possible changes to your bathroom.

It also helps you see that you can do many things with the bathroom you have without ever moving a pipe, an electrical fixture, or even cutting into the wall. You can replace many of the components without major surgery. That not only means less cost, it also means less time. You can improve your home's bathrooms easier than you may have thought.

Facelift Basics

A bathroom facelift is a remodeling project that doesn't require tearing into a wall. You don't have to move plumbing or fixtures. You may be surprised at what you *can* do without opening up a wall. Here are some ideas:

- ◆ Paint or wallpaper the walls
- ◆ Paint or refinish the ceiling

- ◆ Replace the flooring
- ◆ Replace a cabinet
- ◆ Replace or resurface a countertop
- ◆ Replace a sink or faucets
- ◆ Replace lighting fixtures

There are many things you can do to a bathroom without tearing into a wall! You can make a drab or out-of-date bathroom into a decorative and functional room with a brand new look. Even the smallest budget can be stretched to renew a bathroom over a weekend. The first part of this book will help you design and plan the job and the second half will show you how to do it with step-by-step procedures and illustrations.

A facelift with new fixtures, surfaces, and trim can make a dramatic difference in any bathroom.

Resurfacing

The easiest job you can do to remodel and refresh a bathroom is to change the surfaces. It can be as easy as painting or wallpapering vertical surfaces, or move to more complex jobs like installing a new countertop and sink. It can also include installing new flooring materials. You're either recoating or replacing primary surfaces in the room at relatively low cost.

I'll show you how to resurface bathrooms in Chapters 16 and 17. Ideas and tips are also offered throughout this book. Chapters 10, 11, and 12, for example, show you specifically how to remodel half, full, and extended bathrooms. For now, consider that maybe all your bathroom needs is some resurfacing.

Replacing a fiberglass shower enclosure with tile adds a richness.

Paint, wallpaper, and a few new fixtures can make a dramatic change in a bathroom.

Easy Fixture Replacement

Bathroom fixtures are the final components in various systems. In plumbing systems they are the toilet, shower head, tub faucet, and sink faucets. In electrical systems they are the outlets, switches, and lighting fixtures. Because bathroom fixtures are right out in the open (you don't have to tear into a wall to get to them) they are great candidates for replacement.

Bathroom toilets come in numerous colors and shapes to match your decor.

Waste Lines _____

Although it's a cliché, "Don't bite off more than you can chew" is still good advice. Thoroughly consider a bathroom facelift before deciding to a more extensive remodel. You can save thousands of dollars and lots of time with a well-planned bathroom facelift—and probably get the results you want.

Some new fixtures and wall covering can easily dress up a small bathroom.

If you have other storage, consider replacing a cabinet with an elegant pedestal sink.

Even if they still operate well, fixtures can be updated with new designs to make your bathroom look fresher at low cost. You can buy new faucets, for example, for less than $50 each and install them in an hour. A light fixture can be

replaced for under $100 and is only an hour's work. A ventilation fan can be replaced for about the same money and effort. You can replace worn or damaged light switches and receptacle covers for an investment of just a few dollars and a few minutes. You'll learn how in coming chapters.

A Little Extra Effort

Sometimes a tired bathroom needs a little more work. Cabinets become worn or damaged. Towel racks and mirrors need to be replaced. A dented door could be replaced. There's a little more work—and cost—involved in making these changes, but the results are even more dramatic. New or expanded cabinetry can turn an old bathroom into a new one, especially if combined with other facelift projects.

For many readers, all you'll need to give your bathrooms new life is a facelift—replacing, resurfacing, or refreshing components and surfaces. The investment in time and money is relatively low and the results can be amazing.

This bathroom looks totally different with a new cabinet, fixtures, and mirror.

This open cabinet looks freestanding, but actually is attached to the wall and plumbing.

Popular Facelift Projects

The majority of bathroom remodels—especially those done by do-it-yourselfers—are facelift projects as just described. They include resurfacing or replacing, often completed in a weekend or two for a few hundred to a few thousand dollars.

You'll be analyzing and planning your own bathroom facelift in coming chapters. But you need to know what's possible and what's popular. So let's take a closer look at some common facelift projects.

Half Bath Facelifts

In home construction, a *half bath* or partial bathroom is one without a bathtub or shower. It typically includes a toilet and a sink. That's it! Half baths are popular where space is at a premium, as backdoor bathrooms ("Clean up before you come in here!") often located near the kitchen or entryway, and as guest bathrooms when you don't want guests to stay long. Half baths are sometimes euphemistically called powder rooms.

> **Bathroom Words**
>
> **Half baths** typically are less than 25 sq. ft. (square feet) in size. For example, a half bath may be 4 ft. × 5 ft. or 3 ft. × 7 ft. in size. It can be larger, but usually doesn't contain more than a toilet and a sink unless it also serves as a storage area.

Remodeling a half bath is relatively easy. The most difficult part can be maneuvering inside such a small area. If the sink cabinet is wall-to-wall, replacing it may require removing adjacent door trim or other components. The plus side is that there is less surface to paint. Chapter 10 thoroughly covers remodeling half baths.

Full Bath Facelifts

A *full bath* includes a toilet, sink, and bathtub. Sometimes a bathroom with a shower instead of a tub is called a three-quarter bath. Both types of bathrooms typically are adjacent to bedrooms. Some are dedicated to and only accessed from one bathroom, while others are in a hallway to be shared among multiple bedrooms. A shared bathroom also can be one that has door access from each bedroom rather than through a common hallway.

> **Bathroom Words**
>
> **Full baths** are larger than half baths, with a minimum size of 35 sq. ft. More typical is 50 to 100 sq. ft. in size. Larger is better.

This bathroom was enhanced with a simple stained glass window and some new fixtures.

Common full bath facelifts include replacing cabinets and countertops, resurfacing or retiling bathtub or shower walls, replacing lighting fixtures, upgrading flooring, and, of course, painting or wallpapering. Chapter 11 will guide you through specific remodeling projects on existing three-quarter and full baths.

Extended Bath Facelifts

An *extended bath* is one that goes beyond the toilet-lavatory-tub/shower combination, primarily for relaxation. That includes a hot tub, spa, soaking or whirlpool tub, or other amenities. A master bath typically adds a shower and a second lavatory sink. If you want to add one of these luxuries to an existing bathroom you'll learn how in Chapter 12. For now we're concentrating on popular facelifts to existing extended baths.

Bathroom Words

Most **extended baths** are at least 100 sq. ft. in size (10 ft. × 10 ft.) and go up from there. A garden bath, for example, with a double-sink lavatory, separated toilet area, whirlpool, and maybe a decorative cascading waterfall or indoor plant area must be at least 150 sq. ft. in size to accommodate all the amenities.

An extended bath is bigger of necessity. You can't have a hot tub in a small bathroom. Fortunately, extended bathrooms offer a larger pallet for easy remodeling. You don't have to move fixtures, just replace them. You can paint or wallpaper in creative ways. You have room to add decorative touches that change the mood of the room, making it more relaxing and comfortable. You have more options—and higher costs.

You might be able to add a half bath to the laundry room.

Considering Your Options

In fact, you have *many* options. That's what this book is about: presenting and helping you implement numerous options that enhance the livability of your home. One of the first options you have is the bathroom facelift, either as a long-term or short-term solution to decorative functionality. You may decide all that your home's bathrooms need is some surface work. Or you may determine as you get farther into the process that you *really* want an additional or extended bathroom but you'll *settle* for a redecorated bathroom for now—until you can scare up some more funds. All of these options are good ones. I'll show you how to choose the *best* option for your lifestyle, skill level, and budget.

Toward that goal, let's consider why a bathroom facelift makes good sense. It's the simplest option when remodeling your home's bathrooms. It may also be the best.

Minor Changes Can Make a Major Difference

If you haven't already tried this, here's a way to help decide whether a facelift is your bathroom's best option. Go get a couple of cardboard boxes and load everything from your bathroom in them that isn't nailed down. That includes anything on the lavatory countertop, linens, bath rugs, wall hangings, and whatever. If a mirror is hung (rather than mounted), remove it.

Next, leave the room without stopping to analyze what it now looks like. Take a break before returning. In a few hours, slowly walk through your house to remind yourself of its personality or theme (or what you'd like it to be). Finally, enter the cleared-out bathroom and use your imagination to project the home's mood into this room. If you don't think your home has a personality, or you're trying to

change it, stand in the bathroom and consider what touches might make it more comfortable and relaxing. Think of what colors make you feel more relaxed.

> **Heads Up!**
>
> Before starting any bathroom remodeling project, know where the main and branch water disconnect valves are located in your home. Just as you shouldn't work on electrical wiring with the power on, you shouldn't work on plumbing until you're certain that the pressurized water line is closed and water removed.

Finally, make lots of notes. Then take some photos as long as they can truthfully capture the room's current colors and lighting. This is your canvas, your starting point. You now can begin building and implementing a vision, a design, that draws out the desired moods. You may want a cheery bathroom that is primarily used in the mornings for the day's preparations. Or you may want art deco for a guest bathroom to impress others with your sense of color. Or you may opt for easy-to-clean surfaces in neutral colors for the children's bathroom.

A shower shelf adds functionality.

Remember, some paint and wallpaper, maybe a new sink or cabinet, and some decorative touches may be all your bathroom needs to give it a new life—at least until you can afford a major remodel.

Not Much Money

Another good reason to consider a facelift over a full remodel is money. While a total bathroom remodeling job can cost four figures, a quality facelift can be done for less than three figures. Even if you have the money for a full remodel, you may be surprised at the value you get from a simpler touch. In fact, you can use the money you save to hire someone to do all the work while you're off on vacation!

It's not about being cheap. It's about value—about getting your money's worth from every dollar you spend. That means taking the time to consider all your options, making plans that reflect your needs and wants, and having some fun planning and remodeling your home's bathrooms.

Not Much Time

Each of us has 1,440 minutes in a day, no more and no less. Of course, we all spend those minutes differently based on individual goals, demands, and opportunities. Somehow, remodeling the bathroom slips to the low end of the priority list—just below going to the dentist. Life is already sufficiently busy, making it easy to postpone remodeling jobs.

That's why a bathroom facelift makes good sense. Maybe you can't afford the hours it will take to move or add major bathroom components, but you may be able to get great results by painting and replacing things over one or two weekends. A facelift is a good option that helps satisfy current needs in the shortest time.

So hang in there. Even though this book is about *remodeling* bathrooms, it has many practical guidelines for taking the simpler approach, a bathroom facelift, and spending the saved time and money on things you'd rather do.

The Least You Need to Know

◆ Today's bathrooms share the functionality of their predecessors, but offer many more options for privacy, comfort, and relaxation.

◆ A bathroom facelift doesn't require opening walls and can be accomplished for an investment of less time and money than a full remodel.

◆ All sizes of bathrooms (half, full, and extended) can benefit from popular facelift projects.

◆ A bathroom facelift answers the concerns of too little time or money by making significant changes with simple projects.

In This Chapter

- ◆ Moving big things in your bathroom
- ◆ Smart remodeling ideas and tips
- ◆ Ways you can make your bathroom better
- ◆ Do it yourself or hire a pro?

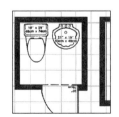

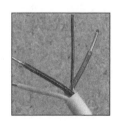

Chapter 2

Bathroom Makeovers

Sometimes simply recovering or replacing bathroom components isn't quite enough. You need to make some major changes. Move a tub. Modify a wall. Change some plumbing in the wall. Can you do it yourself? *Should* you?

This second chapter heaps additional ideas and options on to your innocent plan to modernize a bathroom. It gives you information you may not have considered. And it offers professional tips that can make the job go easier and make a smaller dent in your checking account. As before, don't grab any tools just yet. Instead, let's explore more options.

Makeover Basics

A bathroom makeover changes major components, especially *plumbing*. In most makeover projects you have to access the inside of a wall, floor, or ceiling to move pipes or wiring, called *rough-in* and *finish plumbing*. The work isn't as difficult as it sounds and the results can be amazing. A bathroom makeover can include:

- ◆ Moving or replacing a toilet
- ◆ Moving or replacing a bath tub
- ◆ Moving or replacing a shower
- ◆ Moving or removing a wall within the room
- ◆ Replacing a floor or ceiling
- ◆ Rewiring the room for new electrical fixtures

> **Bathroom Words**
>
> **Rough-in plumbing** is the installation of the basic and hidden components of a plumbing system such as pipes and joints. **Finish plumbing** is the installation of the attractive visible parts of a plumbing system such as plumbing fixtures and faucets.

> **Potty Training**
>
> One of the earliest manufacturers of bathroom fixtures and fittings was Thomas Crapper. Really! In fact, the Thomas Crapper Company has built fixtures for English royalty and commoners for more than 140 years. The American slang word "crapper" was coined during World War I as doughboys used toilets with the manufacturer's name on the door. The British prefer the term "water closet." Many American designers use the acronym "w.c." to designate a bathroom on building plans. Bathrooms are universal.

Of course, once you've moved or replaced one or more of these major components you then have to finish the room off with a facelift (as discussed in Chapter 1). But it's quite doable. Thousands of first-time do-it-yourselfers successfully tackle bathroom makeovers every year. Others choose to do the design and planning, then hire the best pro to do the job. Both save money and benefit from the results for many years to come.

Let's take a quick look at each of the four major bathroom remodeling jobs and how you can do them yourself. Specific illustrated steps are offered in Part 2; the intent here is to help you become more comfortable with the process toward making the decision to do it yourself—or not.

The toilet unit is attached to the waste plumbing using a floor *flange*. It's the mouth of the waste pipe. Two bolts hold the toilet tightly on the flange. In between is a soft gasket to keep the connection from leaking. Other than that, the only connection is the water line that comes in from the wall and connects below the tank. So replacing a toilet simply requires removing the old unit and putting the new one in its place.

> **Bathroom Words**
>
> A **flange** is a connection between two plumbing components such as a toilet and a waste pipe.

Changing Toilets

The modern toilet is relatively efficient and they really don't vary much from model to model. They are standardized. They may operate slightly differently, but their function and process are about the same for all models. That's good. It means that replacing a toilet also is standardized. They aren't tricky to install, just heavy.

A modern toilet only has one or two major components. A two-piece toilet has the bowl (lower) and tank (upper back). A one-piece toilet combines the two into a single unit, making it more efficient but heavier to move. All the flushing mechanisms are inside the tank.

There's more to the job if you are moving the toilet. You must move the drain pipe and the water source pipe. Because the drain and flange are in the floor you probably will have to remove some flooring to make the change. If the water line is moved you'll be tearing into a wall. Again, these are jobs within the capabilities of most do-it-yourselfers (with the instructions offered in Chapter 13), but you need to be aware of what's involved before you decide that it would be nice to have the toilet moved two inches to the left. Considering the work and additional expense, maybe you'll opt to leave it where it is. Or maybe you'll go ahead and move it to an even more convenient spot.

Plumbing runs through wall framing and requires that the wall covering be removed if the plumbing is to be moved.

Here's a look inside a wall of plumbing including framing studs and pipes.

Changing Tubs and Showers

Tubs and showers, too, have fresh water lines (hot and cold) and a drain line. In many cases, you can replace the existing tub or shower without having to move any water or drain lines. The real challenge is getting the old unit out and the new unit in to the room. Fortunately, you can buy a remodel tub or shower that is manufactured to go through a doorway. (New homebuilders often use single-piece tub and enclosure units or shower units, placing them in the room before the walls go up.)

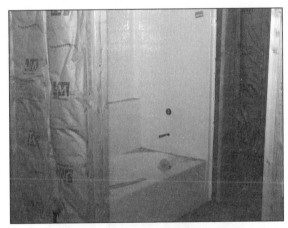

Many single-unit tub enclosures need to be installed before doors are installed.

What becomes more challenging in changing a tub or shower is moving the drain line or converting a tub enclosure into a shower. In these projects the drain and water lines must be moved, necessitating tearing into the floor and wall. Again, the benefits can be well worth the work, but you should know what you're getting into.

Waste Lines _____

Want to do a bathroom makeover, but just can't see yourself moving the drain or water lines? Hire a plumber to help plan the job and even to tackle the toughest parts. You can learn enough from the plumber to do the next tough job yourself.

Structural Changes

Some bathroom makeovers require moving or removing a wall, floor, or ceiling. That's okay; you can do this. Chapter 16 gives you the specifics. In the meantime, know that what seems like a drastic endeavor can make a very welcome change to the beauty and functionality of your bathroom.

For example, you can install a separation wall between the toilet and the other bathroom areas. Or you can add a wall to turn a bathtub into a bath enclosure or a shower. You can add a false floor to raise the new tub (and hide the new plumbing). You can lower some or all of the ceiling for design and functionality.

The corner of this bathroom was modified to hide new plumbing.

Later chapters will get more into the specifics of making structural changes. Also, as you look at the numerous illustrations in this book you will come up with your own unique plans, some of which may require structural changes that you can tackle.

A separation wall offers privacy for the toilet area.

Electrical Changes

Lighting is especially important in bathrooms. In fact, upgrading the lighting is one of the most popular bathroom facelifts. However, sometimes a lighting upgrade requires that you tear into a wall to move or add to wiring and fixtures.

Chapter 14 thoroughly covers remodeling electrical wiring and lighting systems. You'll learn what you can (most everything) and cannot do (very little) and how to do it yourself. The point here is to let your imagination roam free when considering bathroom makeovers. Don't be confined by what is; know that you can make numerous electrical changes yourself. I'll show you how.

Remodeling Basics

I'm a licensed home improvement contractor, but I started exactly where you are, making changes to my own house. I added rooms, remodeled others, replumbed, and rewired. Along the way I learned some very useful tips that have helped me in my professional career as well as in the ongoing jobs that all homeowners face.

This book shares them with you, along with the technical knowledge I've since learned. Here are some of the most important ones, tips that can make your bathroom remodeling—and any other remodeling—project go easier and cost less.

Planning the Remodel

It's the remodeler's mantra: Measure twice, cut once. It means that planning is vital to successful remodeling. Yet too many do-it-yourselfers jump into the project without really thinking it through—and make costly mistakes. You, obviously, aren't one of them because you've purchased this book to see what's involved in bathroom remodeling before tearing into walls. So here are some planning tips that can help you get the job done:

◆ Buy a notebook and start making notes and drawings on your remodeling project.

◆ Go shopping at area home improvement stores, plumbing supply stores, and bathroom shops for ideas and resources.

◆ Start looking at bathrooms, considering what you like and don't like about each, to find out what best fits your needs.

◆ Take a look through this book for additional design ideas as well as a peek at what it will take to remodel your bathroom.

◆ Think what this remodeled bathroom will add to the value of your home, both in terms of comfort and in the eventual resale price of your home.

◆ Ask friends, neighbors, relatives, and co-workers if they have any recent experience with bathroom remodeling.

Chapters 4 and 5 will help you put your initial notes and ideas to work toward remodeling your bathroom. Don't lose that notebook!

Removing Components

Building a home from scratch can be easier than remodeling one. That's because there's nothing to remove. Put up a wall, run the wiring and plumbing, cover the wall, install the fixtures, put on the finishing touches, and you're done. Remodeling an existing bathroom can mean removing the old stuff first: a wall, fixtures, cabinets, trim. Here are some tips on prepping for your remodel:

◆ Make a plan for your remodel prep, listing in your notebook exactly what must be removed and when.

◆ Work safely by making sure water and electricity are disconnected to the room you're remodeling.

◆ Dress for safety, with eye and breathing protection for when dust flies, and appropriate clothing.

◆ Hire or enlist help with tougher jobs like debris removal.

◆ Don't be afraid to ask for professional help if you get into a bind.

Potty Training

Like to know how new homes are built? Read my popular book, *The Complete Idiot's Guide to Building Your Own Home* (Alpha Books, 2002), available through bookstores and online at www.MulliganBooks.com.

Preparing the Bathroom

Work goes much faster and easier if you're ready. More important, you need to make sure that the tools and materials you need are ready to go to work. Here are some tips on preparing for the job to go easier:

- Make a materials list of the components you'll need before you start the job.

- Take your materials list to local suppliers and ask for bids to save time and money.

- Plan your job so that materials show up when you need them, not weeks before or after you need them.

- List and either buy or rent the tools you need to work efficiently.

- Always wear safety equipment and work in a safe manner when remodeling.

Heads Up!

When remodeling, make sure you have a partner nearby who you can call if you need help. Working alone can be dangerous.

I'll offer you more specifics on preparation for remodeling in Chapter 9. You'll also learn about building permits and whether you need one for your bathroom remodel.

Installing New Fixtures

Once you've planned, removed, and prepared for the remodeling job the installation will be relatively easy. Just follow the instructions in this book and those from the manufacturer. Here are some additional installation tips:

- Review the instructions just before making the installation and get any questions you have answered before proceeding.

- Take your time, allowing yourself the opportunity to do the best job you can.

- Don't worry if you make a mistake; most are easily correctable.

- Call for help as needed.

Of course, there are many more tips coming as you begin installing the various components of your bathroom remodel. For now, remember to plan well and work safely.

This cabinet was custom built to match the new décor while fitting into a limited space.

Another unique cabinet that enhances the entire design of the remodeled bathroom.

Popular Makeover Projects

Hopefully you now feel confident that you can at least consider bathroom remodeling projects that require moving big things. If not, keep reading as I describe some of the more popular bathroom makeovers among do-it-yourselfers.

Keep your bathroom remodel notebook nearby. You'll probably see a number of ideas that you can apply when remodeling your throne room.

Half Bath Makeovers

Half bathrooms are smaller and it may seem like there aren't many options. Think again! Here are some idea starters:

◆ Add a window or skylight.

◆ Install a new lighting and ventilation system.

◆ Replace the old toilet with a water-efficient model.

◆ If possible, lift the ceiling for a more open feeling.

◆ Replace damaged *subflooring* around the toilet.

> **Bathroom Words**
>
> A **subfloor** is a horizontal layer of plywood or oriented strand board (OSB) attached to the joists. The finish floor is laid over the subfloor. The subfloor also can be made of concrete.

◆ Replace a standard door with a bifold or sliding door to make access easier.

◆ Modify the fixture locations so the room is easier for the disabled to use.

The subfloor around a bathroom fixture can be repaired or replaced by the typical do-it-yourselfer.

Bathrooms used by the disabled should have adequate spacing, fixtures, and accessories such as grab bars.

Of course, you also can expand a half bath into a full bath by adding a tub or shower, but that means moving some walls. That's covered under bathroom additions and expansions in Chapter 3.

Even a small bathroom can be upgraded with a new cabinet in the corner.

Full Bath Makeovers

There's more room to maneuver—and to remodel—in full bathrooms. They include a

lavatory, toilet, and tub and/or shower. (Some purists call bathrooms with a shower and not a tub a three-quarter bath.) Here are some typical full bath makeover projects:

- Replace an old, worn bathtub with a new designer model.
- Replace an old tub with one that is modern but looks like an even older tub (such as a claw-foot tub).
- Replace a bathtub with a shower enclosure.
- Replace a bathtub with a tub/shower enclosure.
- Modify the bathroom into a garden bath with a plant area, skylight, and maybe a waterfall.
- Add a short or full privacy wall between the toilet and lavatory.
- Add a window or even a door for view or access to an attached garden.

As you look at full bathroom ideas in this book, in magazines, and in others' homes you'll see even more ideas. Make sure you get them down in your bathroom remodel notebook. If possible, get a picture or draw a sketch as well.

Extended Bath Makeovers

If your bathroom is already an extended version—one with a hot tub, spa, or other relaxing amenity—where do you go next? Because you have more room to work with, typically at least 100 sq. ft., you have greater options. Here are some:

- Replace an existing shower head with a wall of shower heads.
- Remodel a larger bathroom into a children's bath to make bath time safer and sized for smaller people.
- Make the bathroom more compartmentalized with a separating door between the lavatory and the toilet.
- Remodel the bath for use by the disabled.

- Upgrade a plastic tub or shower with tile.
- Install a heated floor for comfort on cold mornings.
- Replace an old bathtub or shower with a hot tub, whirlpool, sauna, or other relaxing amenity.

Bathroom Words

There are many types of specialty showers. A **Vichy shower** is a horizontal multi-spray rain bath. A **Swiss shower** is an invigorating vertical multi-stream shower. A **Scots hose therapy treatment** uses pressurized water as massage.

A steam room or sauna can be a welcome addition to an extended bathroom.

- Move the toilet, lavatory, or tub to more convenient locations within the room.
- Enlarge the shower to include a seating area.

There are many more ideas you can consider when remodeling an extended bathroom. This list is just an idea starter. So are the photos in this chapter. Take your time and design the bathroom that best fits your lifestyle and your budget.

You Can Do It!

Your bathroom remodel notebook may be getting full as you jot down lots of ideas, tips, and questions. The biggest question you may be asking yourself, though, is: Can I do this?

The answer is: Probably. Unless you work in the building trade (and you already know this stuff) you will be stretching your knowledge and skills. That's good. And, as a backup, you can always hire a remodeling contractor to advise you and even pitch in on the tougher jobs. Keep reading and you'll discover what you need to know for the job.

Hopefully you'll allow yourself enough time to take your time. You'll read this book through more than once, making copious notes, then use local resources and the resources in the Appendix B to draw up a remodeling plan. Chapters 4 through 9 will be especially important to the planning stage, showing you how to envision the best design and how to make it happen.

One last point: Don't be afraid of making errors. You will! Hopefully, they will be small and easily reversed. Even if they aren't, nothing is perfect. That's why there's paint and trim, to cover problems. Take your time and enjoy what you are doing.

The Least You Need to Know

- Bathroom makeovers involve removing or replacing major components.
- Develop a remodeling plan to make the job easier with fewer problems and lower costs.
- Consider a variety of design ideas before selecting the ones that best fit your lifestyle and your budget.
- Don't be afraid of making and learning from mistakes.

In This Chapter

◆ Adding a new bathroom to your home

◆ Figuring out how to make the addition

◆ Making it happen with less hassle

◆ Looking around for ideas

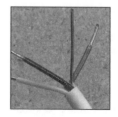

Bathroom Additions

Sometimes the problem isn't remodeling the bathroom you have, but adding on a new bathroom. Now we're talking some serious work—or so it might seem.

Actually, adding a bathroom to your home may be easier than other options. There may already be extra space for the bathroom in a bedroom, basement, attic, or other area. Or you may need to create some additional space. In any case, an addition can be your best option.

This chapter offers dozens of practical ideas and lots of tips on how to enhance the livability of your home with the addition of a new bathroom. You'll see how it's done and you'll see the results. Along the way you'll get some inspiration for your own bathroom remodeling project. Let's get started!

Adding a Bath

A bathroom addition doesn't just remodel a room, it remodels the house. It either turns space previously used for something else—or unused—into a bathroom, or it actually adds a new room to the home.

Adding a new bathroom can be more complex, depending on what's already there. That is, you may be able to add a new bathroom on the floor above an existing bathroom and extend the plumbing and electrical service. Or you might need to extend the home's foundation to accommodate a brand new room before you even start on the walls, plumbing, electrical, and other stuff. A third option is to expand an existing bathroom, turning an adjacent closet into a shower or taking a few feet out of one room to add it to another.

What's involved in adding a bathroom to an existing home? Should you attempt this yourself or hire a contractor? Good questions. Let's discuss them.

> **Potty Training**
>
> Basement bathrooms can be unique, requiring that plumbing flows uphill. I'll cover this topic in later chapters. You also may want to read *The Complete Idiot's Guide to Finishing Your Basement Illustrated* (Alpha Books, 2003) for additional remodeling ideas. It's available at bookstores and online at www.MulliganBooks.com.

Expanding Space

If all you need is a larger bathroom, expanding an existing bath might be the better option. It still requires that major walls get moved and plumbing might be rerouted, but it saves the labor and expense of building a new room. The expansion can go into an existing room or it can move an exterior wall for additional space. Here are some ideas to get your creative juices flowing:

◆ Install a door through an exterior wall to access a patio, where a hot tub or spa can be installed.

◆ Add a small sunroom on the exterior that can become a garden room for your bath.

◆ Reduce the size of an adjacent closet (with storage projects) to gain additional space for bathroom fixtures.

◆ Go up: Cut into the attic ceiling above to install a skylight.

◆ Go down: Change the flooring system so you can install a bathtub with the top at floor level.

Moving a closet gave the owners room for a new shower.

A skylight can add natural light with privacy.

Use your imagination to picture your bathroom expansion options. I'll give you the details on how to do it in Part 2 of this book.

A new alcove can house a new lavatory.

Adding Space

Sometimes there's just no getting around it; you're going to have to add on. How much? Where? What's the cost? Is it really worth it? You'll find the answers to these and many other questions as you read through this book. Meantime, let's consider some of your options if you choose to add on:

◆ Build a new second floor bathroom above an existing first floor bathroom by installing a new dormer (roof addition).

◆ Add a bedroom or other room to your home and convert or expand the old bedroom into a large luxury bathroom, or his-and-her bathrooms.

◆ Move a utility room to the garage and convert the old room into a new bathroom.

◆ Add a utility bathroom in an attached garage and close to existing plumbing systems.

◆ Find a location where the existing house foundation can easily be added to.

◆ Ask a builder for advice on your options for adding space for a new bathroom; hire the builder for the addition and you install the new bathroom.

Get the picture? You have lots of options before you actually have to extend the house foundation. And even if you do extend it you still have options.

Addition Basics

Adding on to your existing residence requires more than just pushing a wall out or extending plumbing and electrical services. You need to make sure the new room has support (foundation) and offers support to whatever's above it (rooms, roof). For that you need to understand how homes are built. Here's a review.

Potty Training

Want to learn more about building a house from the ground up? Get a copy of my book, *The Complete Idiot's Guide to Building Your Own Home* (Alpha Books, 2002) for start-to-finish specifics. It's available at bookstores and online at www.MulliganBooks.com.

Foundations and Floors

Houses are heavy. They need something to hold up the house, protect the people and stuff in it, and withstand a wind storm or other outside pressures. They need a good *foundation.* Most modern residences have well-engineered foundations of reinforced concrete or concrete blocks. Foundations even have their own foundation component, called footings, that are wider to disperse the heavy load.

> **Bathroom Words** _____
>
> A **foundation** is the supporting por-
> tion of a structure below the first-floor
> construction or below grade, including the
> footings.

On top of the foundation of most homes are lumber pieces called sill plates that attach the wood flooring system to the foundation. The sill plates are actually bolted on to the foundation so a strong wind won't literally knock the house off its foundation. On top of the sill plates are joists, the floor's frame system that absorbs all of the house's weight and delivers it to the foundation. The flooring system is topped by wood or plywood called the subfloor.

Walls and Ceilings

The walls are wood or steel frames that sit on top of (and are attached to) the subfloor. The horizontal parts of the frame are called the plates and the vertical components are studs. Studs typically are installed 16 inches apart. Frames are made within walls to hold future doors and windows. The member that's above a door or window to support weight above it is called the header. You'll learn more about walls in Chapter 16.

The ceiling is built similar to the floor. Above it is the roof system, designed to hold the house together and to stand up to local weather conditions. Before covering up the wall and ceiling frames with drywall, plaster, or other materials, the plumbing and electrical systems are installed, described in the next section.

Walls, ceilings, and floors also get _insulation_ to help retain expensive heated or cooled air inside the home. The insulation is installed before the drywall or other wall covering.

> **Bathroom Words** _____
>
> **Insulation** is any material that resists
> the conduction of heat, sound, or
> electricity. Most commonly, insulation refers
> to products installed to reduce the flow of
> heated air in or out of a home or specific
> room.

The ceiling for this addition was raised to give the room an open feeling.

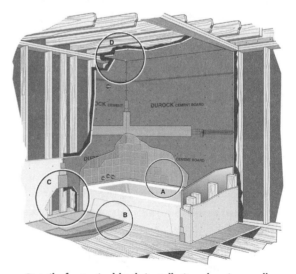

Detail of a typical bath installation showing wall interiors.

Adding a window to this room adds light and beauty.

Plumbing and Electrical

The water and electricity that runs through your house does so inside walls and floors. So once the walls, floors, and ceilings are framed, the plumbers and electricians do their work.

Starting at the point where water is delivered to the house, typically called the main, the plumber runs pipe to the various locations based on the home's building or remodeling plan. Pipe is cut as needed and joined together with fittings. At the end of the pipe will be controls such as faucets, installed later. I'll show you how to install plumbing in Chapters 13 and fixtures in Chapter 18.

The electrician starts where the electricity is delivered to the house, at the *service panel*, running wires to the new receptacles, fixtures, and switches. (Actually, electricians "pull" wire from the receptacle end to the panel for convenience, but the result is the same.) Fixtures are installed before or after the wall is enclosed, depending on the type of fixture. You'll learn more about wiring and lighting systems in Chapter 14.

Bathroom Words

A **service panel** is the box or panel from which the electricity is distributed to the house circuits. It contains the circuit breakers and, usually, the main disconnect switch.

Flooring material is installed over the subfloor.

Additional lighting can make the bathroom even more practical.

Other Bathroom Systems

As you know, there's more to a bathroom than water and electricity. It takes additional systems to keep it clean and comfy, including heat, ventilation, cabinets, and other fixtures.

If you're adding rather than remodeling a bathroom you especially need to know about heating and *ventilation* systems. Heat for the room can be supplied from the existing home heating source, a furnace for example. Or you can install an independent heating source in the bathroom, such as a wall heater. The choice will depend on whether your home's heating system is already overtaxed and how easy it will be to extend the existing system to the new bathroom.

Bathroom Words

A **vent** is any opening, usually covered with screen or louvers, made to allow the circulation of air, usually into an attic or crawlspace. A **ventilation system** typically includes a fan.

Ventilation systems are important to bathrooms, too. Besides renewing the air, the vent system should remove moisture from the air to keep mirrors and fixtures from collecting condensation. Excess moisture can damage surfaces as well as make some flooring materials slippery and unsafe. Chapter 15 will cover bathroom heating and ventilation systems in greater detail. For now, consider that all these systems—foundation, flooring, walls, ceiling, plumbing, electrical, and heating and ventilation—are important bathroom additions.

Popular Additions

As you can see, adding on a bathroom isn't as problematic as you might have imagined. It's a more complex job than a simple bathroom facelift or makeover, but it can be just as successful. In fact, thousands of bathroom additions are being made right now throughout the United States and Canada, many of them by do-it-yourselfers like you.

Pull out your bathroom remodel notebook and make notes as the more popular additions are covered. Some won't work for your home's needs, but others will.

Half Bath Additions

A half bath typically includes a toilet and sink, but no bathtub or shower. You probably don't want to go to the expense and trouble of extending your home's foundation just to add a 30 sq. ft. half bathroom, but you can make it part of a larger remodeling job such as adding a new bedroom and half bath. You can use this opportunity to make even greater changes in livability.

Otherwise, consider moving an interior wall to make way for a new half bath. Here are some of the locations that homeowners select for this project:

- In the corner of a large bedroom
- In a large closet
- In a large utility or laundry room
- At the edge of an attached garage
- Underneath a staircase
- As part of an enclosed porch that will be remodeled into a new bedroom with half bath
- In an attic with a new roof dormer
- In a basement

The space for this half bath is well decorated.

Chapter 10 will give you even more specifics on remodeling and adding on half baths. For now, just consider the possibilities.

Full Bath Additions

A full bathroom includes a toilet, lavatory or sink, and a bathtub or shower. Full baths typically are 50 to 100 sq. ft. in size, though they can be larger or even smaller. Because of their size, about twice that of a half bath, they are good candidates for room additions with an extended foundation. Of course, if you can find a way of adding on a full bath without the cost of extending the foundation you are time and money ahead.

Where should you consider adding on a full bathroom? The more common locations include:

- Adjacent to a main floor bedroom
- Convenient to a group of main floor bedrooms
- Attached to a new bedroom addition
- On the exterior side of an existing bathroom
- In a newly remodeled living space that previously served as a garage
- In a new wing of the home designed to expand living space

This bathroom was expanded to include a separate shower and bathtub.

Full bathroom additions and remodels are covered in Part 2, especially Chapter 11. I'll cover preparation and building permits in Chapter 9.

Potty Training

Remodeling bathrooms can be an investment in your home. If you are investing big bucks on your remodeling job, consider hiring a residential appraiser who can tell you how much your home will increase in value—or not—based on your remodeling plans.

Extended Bath Additions

An extended bathroom is at least 100 sq. ft. and includes fixtures beyond the basics. Besides a toilet, lavatory, and tub or shower, it may have a spa, hot tub, sauna, steam room, special shower, larger fixtures, a dressing or powder room, a garden area or fountain, or other components that make the room for relaxing and longer visits.

Should you consider adding on to your home to have one or more of these amenities? The answer depends on your needs (see Chapter 4) and your budget (see Chapter 6). Meantime, here are some ways of justifying adding on an extended bathroom:

◆ Install a waterproof computer and work from home. (Just don't work from the bathtub!)

◆ Use a second tub as a koi pond.

◆ Install an exterior entrance and rent it out.

◆ Install a hammock, tropical plants, and sand to live out your dream of owning a tropical island that's within commuting distance of your job.

Seriously, there are good reasons for adding an extended bathroom on to your home, but they are subjective. Come up with your own and I'll show you how to make it happen.

This extended bath is part of a bedroom–bathroom addition to an existing home.

Planning Your Addition

Obviously there's more work to adding on a bathroom than doing a facelift or makeover. However, the effort may be a good investment for your home and your lifestyle. For now, don't limit yourself to what's already in place. Continue considering an addition as an option. Chapter 6 on budgeting the job will help you decide whether it can be a reality or will continue as a project for your I-Won-the-Lottery list.

Because additions are more complex projects, you may want to hire some experts to help out. There are architects, bathroom designers, general and specialty contractors, suppliers, and many other folks standing by to give you a hand—with their hand out.

Working with an Architect

Do you really need an architect to help you design a bathroom addition? Maybe. The answer depends on the complexity of the addition and how comfortable you are with doing your own design work. I'll give you more tips on hiring architects and other experts in Chapter 8, but you should begin considering the option now as you think about adding a bathroom.

An architect is someone trained and experienced in the design of structures. That's pretty broad, so you want to work with an architect who is experienced with bathroom additions and remodeling if possible. Some states license architects and others don't, but many are members of the American Institute of Architects (A.I.A.). Others are certified by the National Kitchen & Bath Association (N.K.B.A.).

Architects can be paid by the hour or by the job. The job price can be a flat fee or a percentage of the total value of the remodel, typically 8 to 15 percent. If the job is small, there may be a minimum fee. Good advice is to ask for an initial consultation, either free or an hourly fee, to determine whether you want to use an architect and, specifically, a particular one. Again, I'll offer more guidance in Chapter 8.

Working with a Contractor

Do you need a contractor to make the addition, move walls, or install larger components? Same answer: Maybe. How experienced are you at working with plumbing and electrical systems? Can you read and follow printed plans for the remodel? Do you have the time?

> **Waste Lines**
>
> If you're not in a hurry you sometimes can save as much as 10 percent on your bathroom remodel by hiring a reputable contractor who agrees to fit your project around other jobs. Just make sure you have a firm deadline for completing the job and a penalty for not meeting it.

There are all types of contractors, some licensed and some not. However, many states require that anyone you hire to build or remodel a residence have a license. Most licenses are earned with a specific number of years as a journeyman in the trade plus passing a test and paying a fee. In addition, there are different types of contractors. A general building or remodeling contractor typically oversees other tradesmen, called subcontractors, in doing their jobs: plumbing, electrical, heating, and so on. The subcontractors also should be licensed for their specialty. Most general contractors also do at least some of their own work. Smaller remodeling companies are one person who can do it all.

I'll cover hiring the best contractor for the job in Chapter 8. For now, keep your mind open to hiring a pro to do the job for you.

Doing It Yourself

If you're handy you probably can do a bathroom addition or remodeling project yourself. I'll show you how in coming chapters, especially those in Part 2 of this book. You have many options:

♦ Take some vacation time and do it all yourself.

♦ Plan the remodel for weekends, evenings, or other time off.

♦ Do the remodel with a partner, sharing the work load as time and skills allow.

♦ Hire a contractor to handle the tougher jobs like installing a foundation and framing the addition while you plan to do the plumbing and electrical work.

- Be your own contractor, hiring subcontractors to do their respective jobs and save some money.
- Hire a contractor to oversee your work, giving you one or two days a week of time as your mentor.

You have lots of options. For now, start a list in your notebook of do-it-yourself jobs you've done and what you learned from them. It can help you analyze whether you want to tackle this job on your own or not as well as what specialists you might call on to help. Remember: A half-finished bathroom is useless!

Getting More Ideas

These first three chapters have been about ideas, getting you thinking on whether the bathroom(s) in your home need a facelift, a makeover, or a new room. We'll start getting more specific in the next chapter on analyzing your needs. Meantime, keep the creative side of your brain at work looking for options. Here are some suggestions:

- Flip through this book for additional photos and drawings that help you visualize the bathroom you want.
- Visit larger home improvement stores and bathroom shops for ideas.
- As you visit friends ask to use the bathroom and get more ideas on what does and doesn't work.
- Mention to friends and acquaintances that you are considering remodeling a bathroom and ask for advice.
- Take a look at some of the decorating magazines at the store to see more bathroom ideas in action, remembering that these are show pieces and not typical.
- Ask other members of your family—including kids—for ideas and suggestions.

Most important, think of this project as a solution rather than a problem. Enjoy the search!

The Least You Need to Know

- A bathroom addition reuses or expands the living space in your home for a nobler purpose.
- Adding a bathroom to your home may require a new foundation, floor, walls, ceiling, plumbing, electrical, and other systems.
- Consider hiring a contractor or architect to help you add a new bathroom.
- Look for bath design ideas all around you.

In This Chapter

- ◆ Figuring out what you really need in your bathroom
- ◆ Making sure everything is designed to code
- ◆ Taking inventory of what your bathroom offers
- ◆ Inspecting the bathroom for hidden problems

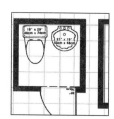

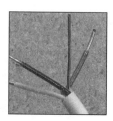

Chapter 4

Analyzing Your Needs

By now you've developed some ideas on what type of bathroom remodel you're going to tackle: facelift, makeover, or addition. Of course, the final decision is subject to the design, budget, and how much you're going to tackle yourself.

This chapter takes a look at your bathroom as a service rather than just another pretty room. It helps you determine your living group's needs, the design rules you'll have to live with, and any "issues" that need addressing as you remodel. It's an important step in getting the room you want at the price you can afford.

Needs Analysis

Let's talk about the functional requirements of your bathroom. What functions does it need to fulfill for your family or living group? Some are obvious and others may not be, but a well-designed bathroom is one that considers all needs and anticipates future possibilities.

A needs analysis is the first step toward remodeling a bathroom that will serve you for many years. It also can save you money as you decide what you *really* need versus what you'd like to have if you find a few extra bucks in the budget.

This ideal master bedroom bath is spacious, well illuminated, and functional.

The Needs Interview

Modern bathrooms are designed for privacy, but they are shared. In some living groups, a single bathroom may serve the needs of many people, including friends and guests. That's a lot for one room to do.

Fortunately, most of those needs are identical or at least similar. Functionally, those who use a bathroom need a toilet for elimination, a lavatory for washing hands, a mirror, and a power source for efforts at personal beauty, and a shower or bath for hygiene. Storage also is needed for service supplies: toilet paper, towels, soap, hair dryer, beauty supplies, and so on. That's the short list. The longer list includes many options that serve additional needs. It also considers use or traffic. Here are some questions to answer as you determine bathroom needs and how to meet them.

- How many people will use the room during a typical day?
- What services do each of these people typically need?
- What services can be easily shared and which not as easily?
- What are the ages of the users?
- What changes do you expect in the needs of users over the next few years?
- Will this room serve as the primary bathroom for the entire house, one area of the house, or just one bedroom?
- Do users need a place to relax, such as in a whirlpool bath or with a stereo system?
- Are group members frequent bathers that will need additional towel storage?
- Do group members prefer total privacy when using any of the room's services or are they willing to share for some? Which?
- Is instant hot water needed for washing and shaving?
- Does bathroom storage require safety locks for children or for personal privacy?
- Is better lighting needed?
- If you could move this bathroom anywhere in the house, where would it be most convenient?

Extra storage space can be installed in otherwise wasted corners of your new bathroom.

Some of these questions might seem silly, but answering them can help you design the best plan the first time. For example, depending on your living group and the number of other bathrooms, a shared bathroom in the hallway may serve as the main bath or shower, but household members may use one of the other bathrooms for the toilet and primping. An elderly group member may need special equipment, easier access, or a bathroom closer to the bedroom. Think of the bathroom first as a service center rather than just a room to redecorate. Make sure it's functional first.

Potty Training

If you share the room with others you should get their inputs. In fact, asking for and considering other viewpoints can help cement relationships. Yes, people do break up over events like bathroom remodels. And through this remodeling effort you may need all the friends you can get, so be sure to ask others to participate in the design.

The Needs List

From these questions you'll come up with a better understanding of what the remodeled bathroom requires to meet the needs of the people it serves. This is your bathroom needs list. Here are some typical needs to help you fill out your list:

◆ Two sinks for personalized hygiene areas

◆ At least four electrical plugs near the mirror for hair and shaving appliances

◆ Handrails around the toilet for easier access by larger or senior family members

◆ A wheelchair-accessible shower

◆ Overflow storage for seasonal household linens

◆ Electrical power for a corner waterfall's pump

◆ A compact water heater for quicker service

◆ A television viewable from the bathtub

◆ An elongated toilet for comfort

◆ A telephone handy to the toilet

◆ A privacy wall between the toilet and lavatory

◆ Relaxing lighting and sound suitable for long soaks

◆ A sauna or steam room

◆ An area for exercise equipment

Okay, so some of these "needs" are really "wants," but they may be necessary to enhance the enjoyment of your life. Why work hard to earn if you can't have some enjoyment from your efforts? One person's luxury is another's necessity!

Prioritizing the List

Of course, you may not be able to have all of the things on your needs/wants list. Your new throne room shouldn't cost more than the house did. I'll help you choose features by priority in Chapter 5 on designing and in Chapter 6 on budgeting. For now, put them in the order in which you feel they are most important to your needs.

For example, additional privacy within the bathroom might be a higher priority for you and your living group than, say, installing a television. Or setting up a comfortable place to relax in the tub and watch sitcoms might be at the top of the list. It's a judgment call—but it's *your* bathroom.

Bathroom Requirements

Before we move on to "what's in *your* bathroom," let's consider some of the requirements that new bathrooms must have. After all, you probably will need a building permit to do this remodel and permits are only issued if the plans comply with local *building codes*. Maybe the bathrooms in your home, built years ago, don't even comply with current building codes. If so, you'll probably have to bring them up to code standards to get approval. (I discuss permits and building codes in detail in Chapter 9.)

> **Bathroom Words**
>
> **Building codes** are a collection of written legal requirements for the construction or remodeling of habitable and related structures.

In addition to codes there are some good-sense design issues recommended by the remodeling industry to make the finished job more livable. Here are some of the code and good-sense requirements for remodeling a bathroom:

- Bathroom doorways should be at least 32 in. wide.
- Walking areas should be at least 36 in. wide.
- Lavatories or vanities typically are 30 to 32 in. high, but can be built up to 36 in. for taller people.
- The center of a sink should be at least 15 in. from the closest wall, and 18 in. for washing hair.
- Dual sinks should be apart at least 30 in. center to center.
- Sinks should be at least 8 in. from a counter edge.

- Allow at least 16 in. from the center of the toilet to any other object.
- Allow a 48 in. × 48 in. space in front of the toilet for access.
- Install the toilet paper holder 26 in. above the floor and 6 in. beyond the front of the seat.
- Make sure the shower is at least 34 in. × 34 in. in size.
- Install the shower door to swing out of the shower and into the room.
- Enclosed toilet areas should be at least 36 in. wide by 66 in. deep.
- Enclosed toilet area doors should swing out of the area and into the room.

Make sure your bathroom plan meets minimum design requirements.

There are other codes and requirements as well, covered as appropriate in Part 2 on the steps of remodeling. You'll also get more design guidelines in Chapter 5 on designing your bathroom.

Bathroom Inspection

Now you know approximately what you *want* in your remodeled bathroom. Let's take a look at what you *have*. If you're adding on a new bathroom this inspection is easier because there is no room yet. If you're doing a facelift or remodel, pull out your bathroom design notebook and let's inspect.

The first thing to include in the inspection part of your notebook is a sketch of the bathroom you are remodeling. Include all walls, doors, cabinets, and fixtures as they are now. You'll redraw the layout on graph paper or with a software program in Chapter 5, then use it as a template for your new design.

Inspecting the Air

Well-ventilated air is important to the comfort and hygiene of your bathroom. Besides odors, bathrooms are susceptible to moisture and the damage it can cause. Moisture can collect on walls, in fixtures, and other surfaces as condensation and make a breeding ground for molds and other unhealthy things. The solution is to minimize moisture in the air by eliminating it from the room with adequate heat and ventilation.

Many bathrooms use a small fan unit in the ceiling to ventilate air. Older bathrooms may not have such a vent, relying on air circulation through the home to remove air from the room. In modern homes with energy-efficient insulation, windows, and doors, there may not be enough ventilation without at least a minimal vent fan and system. Local building code will tell you how much ventilation your new bathroom needs.

Inspect the existing ventilation system, if any, by removing the cover and looking for a power tag that tells the unit's power in watts (W) or displacement in cubic feet per minute (c.f.m.). Visually inspect the vent for obstructions or damage. Most vents don't have filters so dust and hair can attach itself to the fan unit and clog flow. Clean it with a vacuum.

The next part of this inspection isn't as easy, depending on the design of your house. Vent systems must vent air to somewhere. Where?

You'll need to crawl up in an attic or check an exterior wall vent to make sure that the bathroom air really is leaving the building. Alternately, you can hire a house inspector to check your home's vent system and tell you what you're working with.

Heads Up!

Most states don't license home inspectors, so just about anyone can get cards printed up and be in the home inspection business, qualified or not. However, they do have a trade association, American Society of Home Inspectors (932 Lee St., Suite 101, Des Plaines, IL 60016; 1-800-743-2744; www.ashi.com). ASHI has a standards of practice and a code of ethics.

Inspecting the Water

With all the water fixtures in a bathroom it's no wonder that water damage can be a big problem. In fact, it can get so out of hand that the only solution is to rip out nearly everything and replace it—a good excuse for a bathroom remodel.

Your inspection of the existing bathroom continues by checking the condition of walls, floor, and ceiling for water damage. The most obvious places to look are around the base of the toilet, around the front edge and drain of the shower, and around the perimeter of the tub where it meets the wall or floor. You're looking for soft spots. Press on a portion of the wall or floor that is away from these areas so you can determine what they should feel like. Then check these common problem areas by pressing on the surfaces. If there is much of a difference there probably is some damage to underlying materials. Indicate the locations on your bathroom sketch so you can identify them later for repair. What you're doing right now is determining if your remodel will include any repairs.

Is your bathroom getting enough water pressure? If not, there may be additional repairs to a water system that is clogged or leaking and may need repair. How can you tell? Residences with public water typically have a meter so the water company knows how much to charge you. Most of these meters also have a pressure gauge. Though pressure delivered to the house can be as high as 100 pounds per square inch (p.s.i.), they typically run from 20 to 60 p.s.i. That's more than enough to simultaneously run a few faucets in the home and get a strong flow at each. If not, there may be a leak or obstruction. That means more repair before you can add or upgrade a water-guzzling bathroom. Call a plumber to help you trace down and repair problem pipes. They can be damaged by age, heavy equipment, tree roots, or poor workmanship. In any case they will cost money to repair, something to add to your remodeling budget.

Inspecting the Power

We are surrounded by electrical gadgets, and the bathroom seems to be their favorite spot in the house. There are electric hair dryers, hair curlers, shavers, lighted mirrors, vent fans, light fixtures, and more. Will your old bathroom's power system support all these new gadgets?

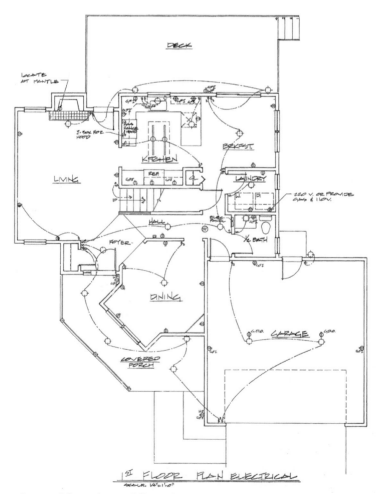

A remodeler's electrical floor plan shows how power is distributed through the house.

The place to start answering this question is at the electrical service panel where electricity comes into your residence and is distributed into various circuits. Each circuit has a *circuit breaker* or *fuse* as a safety limit. All the breakers or fuses typically are together in the service panel. Each breaker or fuse has its limit printed on it: 15A (amp), 20A, and so on. If whatever is plugged into that circuit draws more than its limit of power the breaker or fuse will stop electricity flow to the circuit. Breakers can be reset, but fuses require replacement.

Bathroom Words

A **circuit breaker** is a safety device used to interrupt the flow of power when the electricity exceeds a predetermined amount. A **fuse** is a safety device for electrical circuits that interrupts the flow of current when it exceeds predetermined limits for a specific time period. Two popular types are cartridge and screw-in fuses. Unlike a fuse, you can reset a circuit breaker.

Take a look at the service panel to see if the circuits are labeled as to what circuit they belong: kitchen, upstairs bathroom, living room west wall, and so on. If not, you need to find out. There are handy gadgets you can plug in to a circuit to tell you which of the breakers or fuses in the panel serve it. Most folks, however, turn on all the lights and plug lamps into all the outlets in the bathroom, then turn off one circuit breaker or *carefully* remove one fuse at a time until a helper says the lights went out in the subject room. Then the circuit is marked in the panel. Note that some rooms may have two or even more circuits.

The problem can be subcircuits. Older homes were built with 100A service, but modern homes typically have 200A electrical service. When older homes are remodeled over the years, electricians often install a second service panel. It is often located next to the main service panel, but also could be installed near an addition such as a new bedroom. So you may have to search the house for a second panel to make sure you know exactly what electrical service your home and bathroom are getting. Alternately, an electrician can do this job for you for $100 to $200 depending on the complexity. If you are hiring an electrician anyway, some will do it for free as part of their initial consultation with you.

The electrical service panel is your home's electrical distribution point.

Inspecting the Room

Finally, take a good look at the existing room itself. Determine if any rooms are under it or over it and how the walls match up. It may be that you can easily move water fixtures from one side of the bathroom to the other because of other nearby plumbing. Take some measurements of adjacent rooms to see if there are any hidden pockets that could be used for new pipes or ventilation.

If you are considering putting a skylight in your remodeled bathroom, take a look at the existing ceiling. It may be relatively easy to install a skylight well between roof rafters in the ceiling to allow sunlight into the room. If not, consider a solar tube or a wall window for additional light with privacy. I'll cover installation in Chapter 16.

The more you know about the existing bathroom the better you can design a new one—and you'll stay ahead of costly surprises.

One more point on inspections: Consider hiring a trained and experienced home inspector to take inventory of the bathroom or other area you plan to remodel. The cost typically is $150 to $300. It will not only save you climbing around in the attic or under the house, it will give you more confidence that you know what you're dealing with. Alternately, if you've decided to use a contractor to remodel your bath the contractor will do the initial inspection. Just make sure it gets done.

Waste Lines

If you have friends or relatives who work in the building or remodeling trades—contractors, plumbers, electricians, inspectors—ask them to take a look at your remodeling project to determine if there are any special problems you will face. You can save some time and money!

That's the needs analysis for planning a bathroom remodel. It can be as simple or as complex as any other remodeling job, but it needs to be thorough. Why? Because you will be using it to design your new bathroom as well as determine costs. You don't want surprises, such as opening up a wall to find out that it has extensive water damage that will cost you thousands more than your budget allows. And you don't want to plug the first appliance in to your new bathroom and have it trip a circuit breaker. So even if you need to hire an inspector or contractor to do the job, make sure you get an accurate account of what you're dealing with. Knowledge is power.

The Least You Need to Know

◆ Spending extra time determining what your needs are can save thousands of remodeling dollars.

◆ Unless you have an unlimited budget, prioritize your bathroom remodel needs list.

◆ Make sure you know the requirements for bathroom and fixture sizes and placements.

◆ Perform a thorough inspection of the area that will become your new or remodeled bathroom so you're ready for any costly problems.

In This Chapter

- ◆ Comparing notes with bathroom designers
- ◆ Understanding bathroom design basics
- ◆ Drawing up plans for your bathroom remodel
- ◆ Design elements for the disabled

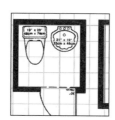

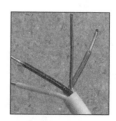

Chapter 5

Designing Your Bathroom

For some do-it-yourselfers, designing a new bathroom is the fun part. For others, it's the most difficult aspect of a bathroom remodel. All those decisions!

Fortunately, you're not the first one to face bathroom design issues. There are lots of experts out there ready and willing to share what they know and to answer your questions. With their help, this chapter will make you feel more comfortable—and even excited—about designing your home's bathroom.

What People Want in Bathrooms

Once a year, the National Kitchen and Bath Association (NKBA) hires a research firm to conduct studies on what people want in bathrooms. The survey asks members what consumers are including in new bathroom designs, so it's skewed toward consumers who use *Certified Bathroom Designers*. However, the results also reflect what many do-it-yourselfers want in their next bathroom.

Potty Training

A **Certified Bathroom Designer** is a certification program managed by the National Kitchen and Bath Association, requiring seven years' design experience including at least three years specifically with bathrooms. In addition, they must take 60 hours of coursework and pass an extensive exam.

Design Elements

Recent NKBA Design Trends Study first notes an increase in remodeling jobs due to a downturn in the stock market. People are investing money in their homes—with greater dividends. Here's a summary of that study:

Bathroom designs can combine traditional elements such as a pedestal sink, with modern touches such as a seating area to make the bathroom more comfortable. Adding color to the walls is another modern touch.

◆ More elaborate and luxurious bathroom styles are increasingly popular as more people stay at home.

◆ Sleeker European designs are preferred for many new bathroom remodels.

European bathrooms include sleeker fixtures and, in some, bidets (shown at left in photo).

◆ Neutral colors are more popular for fixtures and flooring.

◆ Wood cabinets are increasingly popular, especially cherry and maple.

Free-standing bathroom cabinets are increasingly popular for bathroom remodels.

◆ Bathroom faucets are often nickel plated rather than stainless steel to reduce costs.

◆ Granite countertops are very popular.

◆ Cabinets are being designed to look more like furniture.

◆ Simple cabinet hardware is being replaced by decorative designs.

- Showers are becoming more elaborate, with multiple shower heads.
- Entertainment systems are being included in many new and remodeled bathrooms.
- Environmentally friendly products and designs are increasingly popular.
- Many new bath designs include a baby changing room or area.
- Upscale bathrooms/relaxation rooms are including beverage bars with refrigerators, coffee maker, and even microwave ovens.
- Consumers are more knowledgeable about design and price conscious than in the past.

The Budget

How about budgets and costs? How does the typical bathroom remodeling budget break down? The NKBA Design Trends Study says:

- About 21 percent of the typical bathroom remodeling budget goes for new cabinets.
- Bathroom appliances account for 15 percent of the typical budget.
- Countertops typically take up about 11 percent of the bathroom remodeling budget.

- About 9 percent of the budget is spent on fixtures and hardware.
- Flooring takes up about 7 percent of the bathroom remodeling budget.
- Five percent of the typical bathroom budget goes for lighting.
- Five percent of the budget pays for bathroom design services.

The study also noted that about 40 percent of the bathroom designers surveyed don't charge a design fee for their efforts. Those who do typically charge a fee based on the time required. Fewer are paid as a percentage of the job budget or charge a flat fee for all jobs.

I'll cover planning your own bathroom remodeling budget in Chapter 6. Meantime, take notes on your own design ideas.

Designing Bathrooms

The day of the utility bathroom is fading. In its place is a multifunctional bathroom that serves more than just hygiene needs. It is a place to relax, a personal hideout where one can take a long bath or an invigorating shower. A bathroom is no longer a chamber pot and a wash bowl in the corner of a bedroom.

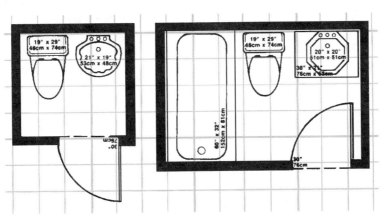

Typical measurements for half baths (left) and full baths (right).

Because you'll be spending more time in the bathroom it makes sense to spend more time designing it. Where to start? With your own tastes. You can hire someone to help you draw them out and put them on paper, but it is *you* who ultimately makes the decisions.

Using a Professional Designer

I'll give you some tips on hiring a professional bathroom designer in Chapter 8. Meantime, here's a summary of what they do.

A designer is a consultant. He or she uses knowledge and experience—and lots of people skills—to help you make decisions on what you want. As mentioned earlier in this chapter, the designer typically is paid a fee or percentage for professional skills, or design services may be included in the pricing. No matter how it is paid, you are paying it. Make sure the design you choose reflects your own tastes, and don't get talked into something you know you ultimately won't like.

There are various trade associations for home remodeling designers. One that I've already mentioned is the National Kitchen and Bath Association. Another is the American Society of Interior Designers (ASID). Many people prefer working with designers who are NKBA members if they are remodeling a bathroom or kitchen, and designers who are ASID members for larger remodeling jobs that encompass more elements of the house. Still other people can get enough design help from an architect or even a design student to make needed decisions.

Potty Training

If you have access to the Internet, check out the National Kitchen and Bath Association at www.nkba.org. The American Society of Interior Designers is online at www.asid.org.

Design Basics

What do you need to know about design to make decisions about your remodeled or new bathroom? You need to know what you like and how to achieve it. Design is the implementation of style, so let's first consider what styles you like.

Styles can be categorized by eras they reflect or where they became popular: colonial, modern, neoclassic, French, Victorian, Arts & Crafts, and so on. If you're not familiar with these styles of furniture and decorating, visit a large furniture or decorating store and ask for an education. Many stores will share information with you because they know that knowledge leads to sales. I'll summarize popular styles in the next section.

Additionally, styles can be categorized by colors, shapes, and textures: earthy, bright, soft, ornate, subdued, and so on. Many people determine what they like in colors and shapes, then select eras or periods that closely match them. For example, preferring pastels and older looking furniture may lead you to pine cabinetry with pastel milk paints accented by simple bronze faucets. If you like simple lines of 1920s furniture, consider Arts & Crafts (also known as Mission) style.

Most important, make sure that the colors, style, and design you select reflect your lifestyle. Are you formal or informal? Do you like softer colors, simpler patterns, and curved edges, or does your taste lean more toward brighter colors, busy patterns, and striking angles? You can carry your home's theme into the bathroom design—or you can do something very different. Hey, it's *your* bathroom!

Design Categories

To help out, here are some general descriptions of popular home designs. Use them as a guide to help you analyze your own preferences as well as relate them to designers:

- Arts & Crafts (Mission): simple geometric patterns, gentle lines, cherry-stained oak, copper, stone tile
- Contemporary: minimalist, high tech, clean lines, soft or bright colors, laminates, chrome, glass
- Country (Farmhouse): pine or oak, painted or stained furniture, handcrafted, patterned tiles, coarse fabrics, nostalgic
- Traditional: classical and elegant with richer woods (mahogany, cherry), porcelain, silver, and strong colors
- Victorian: darker colors, curved woods, overstuffed upholstery, stenciled cabinetry, lace, fringe
- Eclectic: a mix of any or all of the above and whatever else you want

Of course, there are many other design categories including some you can make up yourself. Many people start with a photo, a piece of hardware, a favorite bathroom fixture, and build from there. Take these ideas to a bathroom store

or plumbing retailer to expand your ideas. Talk to friends and family members whose sense of design you admire. Get all the input you can.

Waste Lines

If you get stuck, don't be afraid to ask for help. Once you've narrowed your ideas down you can hire a bathroom designer to help you finalize your design, saving money. The designer can help you determine accent colors or embellishments that can dramatically enhance the design and long-term enjoyment of your remodeled bathroom.

What to Include

Remember to review your bathroom remodel notebook as you begin planning your new bath. Also make sure you discuss ideas with other members of your household. As a guide, don't forget to consider:

- Access from adjacent rooms
- Additional natural lighting
- Better ventilation
- Bidet
- Dressing area
- Fitness area
- Grab bars at toilet, tub, and shower
- Hand-held shower
- Linen and towel storage
- Lounge area
- Makeup mirror
- Medicine cabinet
- Separate shower

A separate shower can be installed in a corner of a room to take advantage of two existing walls.

- Shower only (no bathtub)
- Sit-down vanity
- Sound system
- Steam shower
- Storage options (for hair dryers, shavers, toiletries, and so on)
- Television shelf and cable
- Towel warmer
- Two lavatories
- Upgraded electrical lighting
- Walk-in closet
- Wet bar
- Whirlpool (one or two person)

Make sure you have thoroughly used your bathroom design notebook, filling it with ideas, before moving to the next step: drawing up the plans for your remodeled bathroom.

Drawing Up Plans

Once you made primary design decisions including location, size, shape, components, and colors, you need to find a way to communicate your decisions to others. Who? First, the local building department will want to know what you're up to before they issue a building permit (see Chapter 9). In addition, you'll need written plans to help estimate materials as well as to guide you and any other remodelers on the job. "Let's see, I think I wanted the bathtub somewhere about here."

The answer to keeping everything on track, of course, is to draw up plans. To be valuable, those plans must be proportional and drawn to scale. You don't want to try to fit a 6 ft. bathtub through a 4 ft. opening. There are many ways you can draw up your plans. Let's consider each to determine which works best for you.

Drawing Sketches

Most folks start with sketches in a notebook or on graph paper. In fact, they make *many* sketches with various modifications: the tub on one wall then another, one with a larger lavatory, another with a shower instead of a bath, and so on.

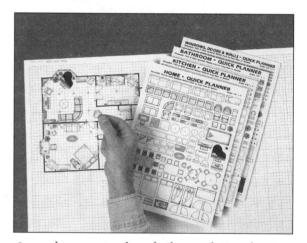

Larger home centers have bathroom design planning sheets like this QuickPlanner from Homeplanner (www.homeplanner.com).

You also can buy design kits through home decorating centers. The kits include graph paper and cutouts of common components that you can lay this way and that, easily moving them around until they fit your ideas. One advantage to these kits is that the components are scaled to size. That is, the toilet and tub cutouts are the correct size to match the graph paper. Some of the labor is done for you.

You'll probably go through numerous sketches and drafts before you finalize on one or two preferred designs, typically one design with a variation. You then can make an architectural drawing of the sketches.

Architectural Drawings

An architectural drawing includes all components, dimensions, and the location of services such as plumbing, electrical, and ventilation. This is the plan that your local building department and even your materials supplier will use. If you are using a remodeling contractor, he or she will need this drawing. In fact, most contractors will help you turn your sketches into architectural drawings.

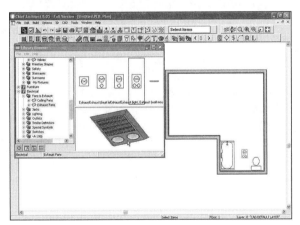

You can use design software to select and plan ventilation units in your remodeled bathroom.

Scale is important to drawing your bathroom plans. The typical scale used is ½ in. = 1 ft.

That is, a 10 ft. × 12 ft. bathroom will be drawn on paper as 5 in. × 6 in. Using graph paper (available at stationery stores) marked out in ½-in. grids, the room will be drawn as 10 grid boxes by 12 grid boxes, 1 grid box per foot. This makes measuring much easier. Write the scale you are using on the drawing somewhere.

Heads Up!

When drawing your bathroom plan to scale, remember to allow for walls. That is, if the interior of the room is 10 ft. × 12 ft., make sure that your drawing shows that as the *interior* dimension and adds wall thicknesses to it.

If you're not sure of the final placement of a few components, make a scale drawing of everything you are sure of then photocopy it and make variations on the copies. Once you have things the way you want them you can revise the main drawing.

Make sure you show the dimensions of all walls and areas on the drawing. If the remodeled bath is 12 ft. long with a short wall in the middle, indicate the total space (12 ft.) as well as the compartments (for example, 8 ft. and 4 ft. less wall thicknesses). Your drawing also should include the location of primary plumbing (water, drain, waste) and electrical services. To avoid confusion, draw the wiring that runs from a switch to a fixture as a curved dotted line.

If you are thorough and careful your own architectural drawing will be adequate for the building department, contractors, and suppliers. If you'd rather not do the final drawing, take it as far as you can and hire an architect to produce the final drawing. You'll learn new things and save money.

Design Software

Another popular option is using architectural design software. If you use computers you know that they are tools that can be powerful or confusing—or both. There are numerous software programs available to help you design houses and rooms. Consumer versions are priced under $100 and vary in features and complexity.

Potty Training

Information about Professional Home Design is available online at www.punchsoftware.com.

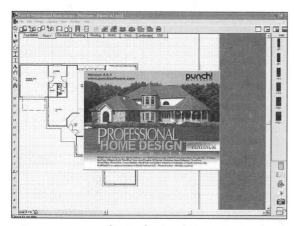

Computer screen for Professional Home Design from Punch Software, popular with consumers.

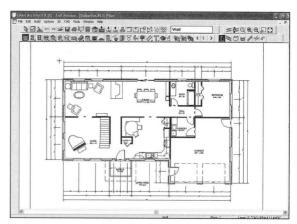

Software programs are available for computers to help you design your new bathroom and other rooms of your home.

Professional Home Design (PHD), distributed by Punch Software, is a popular home construction and remodeling design program that is actually a family of programs. PHD allows you to start a design by quickly drawing the perimeter of the foundation or room, then adding the flooring system, electrical, plumbing, roofing, HVAC (heating, ventilation, and air conditioning), and bathroom fixtures. You can draw your existing home then start remodeling a specific bathroom so you get the big picture, or you can simply draw a floor, walls, doors, and other components until you're done. PHD and many other consumer design programs also can develop and print materials lists for projects. You can use these lists to get bids from suppliers.

One of the most popular components of this and other home design and remodeling software packages is the 3-dimensional or 3D view. Once your plan is completed you can virtually walk through your remodeled bathroom on the computer screen, turning left or right as needed. In addition, many of these programs can instantly strip away the surfaces to expose floor joists, wall studs, and ceiling rafters.

Other consumer home design software includes myHouse from DesignSoft, Classic Home Design from Artifice, SmartDraw, and 3D Home Architect developed by Advanced Relational Technology and distributed by Broderbund Software.

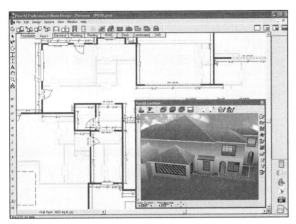

Three-dimensional home design software allows you to see approximately what your remodeled home will look like.

Potty Training _____

Learn more about myHouse online at www.designsoftware.com, Classic Home Design at artifice.com, SmartDraw at www.smartdraw.com, and 3D Home Architect at www.broderbund.com.

On the higher end are products like Chief Architect from Advanced Relational Technology (ART) and sold direct. Developed for architects, builders, designers, and drafters, it is more expensive than consumer design products. But then, you get what you pay for. The library that comes with it includes 7,000 symbols, textures, and images to make a walk-through almost feel like you've broken into someone's home. You can even specify a door or window by manufacturer and size, for example, and plop it into your design to see what it will look like. Once you're done you can have the program print out a one-dimensional model that can be cut and assembled into a three-dimensional bathroom or house. Talk about seeing what you're getting in to!

Potty Training _____

You can find out more about Chief Architect online at www. chiefarchitect.com.

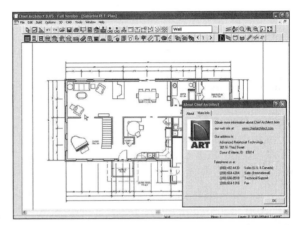

Computer screen for ART Chief Architect professional design software.

There are numerous other features available in consumer- and pro-level home remodeling design software. And there are many other software programs available as well. Larger computer software stores will have a few on hand. Otherwise, check online, download trial versions, and have some fun designing your new bathroom.

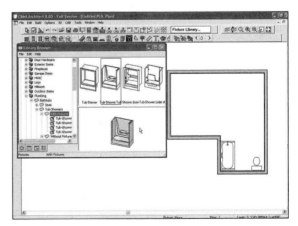

Some design software offers drawings and specifications for actual bathroom fixture models.

Designing for Disabilities

For most consumers, designing for disabilities isn't an issue. However, it should be considered by anyone who is remodeling a bathroom or other home component. In many cases a minor change to plans can enhance accessibility for the disabled as well as the limited. It can be a real benefit if you or someone in your family becomes injured, incapacitated, or limited in movement. A skiing accident can be less limiting if you have a bathroom that accommodates a wheelchair or makes it easier for someone with a cast to get in and out of a tub or on to a toilet. In addition, your home is more saleable if equipped with a bathroom that is accessible by the disabled.

The Americans with Disabilities Act (ADA) of 1990 included many pages of specifications and guidelines on designing for people with physical limitations. If needed you can get a copy of these guidelines through larger bookstores, or online at www.adaproject.org. The guidelines are extensive and cover everything from bathrooms to parking. Here are some basic tips on making your remodeled bathroom more accessible by everyone:

◆ Doorways should be at least 32 in. wide.
◆ Include a clear floor space circle of at least 60 in. diameter for turning a wheelchair for access to toilet and bath.
◆ Install a grab bar 33 in. to 36 in. above the floor on the back and side wall of bath tubs and showers.
◆ Consider installing a roll-in shower with no front edge specifically designed for wheelchair access.
◆ Install grab bars at one side of and behind the toilet per ADA guidelines.
◆ Include lavatory sinks that can be accessed by users in wheelchairs.

There are many other guidelines and specifics, more so for bathrooms that are available to the general public. However, there are things you can do to make your home bathroom more accessible to the disabled by designing them in. Many are relatively inexpensive and can be invaluable to users. Also look at specialized fixtures for the disabled (tubs, showers, toilets, lavatories) as you go shopping for your bathroom remodeling components.

The Least You Need to Know

- Modern trends in bathroom design are toward multipurpose rooms that are relaxing as well as functional.
- Consider your own tastes first as you design your bathroom rather than following the latest fashion.

- You can draw up your own remodeling plans with graph paper or one of the many design software programs available for consumers.
- Include design elements in your remodeled bathroom that make access easier for the disabled and limited.

In This Chapter

- ◆ Making a list of the things you need to do and buy
- ◆ Reality-check: comparing budgets against typical bathroom remodels
- ◆ Figuring what the remodel will actually cost
- ◆ How to get the best value for each dollar you spend

  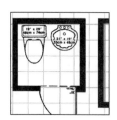  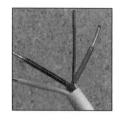 

Chapter 6

Budgeting the Job

Here comes the bottom line, the question you've wanted answered since opening this book: How much is this bathroom remodel going to cost?

This chapter helps you answer that question with specific steps and typical costs. You'll see how to figure all costs as well as how to cut some of them without cutting quality. So sharpen your pencil and let's start spending money—on paper.

Figuring What You Need

The first five chapters of this book have brought you to the point of knowing what you *can* do, what you *need*, and what it will *look* like. Too many people become discouraged by not going through this planning process before hearing a big number and deciding that bathroom remodeling isn't for them. You're tackling the same question the smart way by getting your facts before you get your figures.

What is it that you actually need before you can put a dollar amount on this remodeling project? You need to finalize your design plans, determine what materials are needed, and figure out who's going to do the work.

Finalizing Plans

Chapter 5 guided you through designing your bathroom remodel. If you didn't complete it then, now is the time that you need to finalize those plans. To make an accurate estimate of the costs, you need to know what goes where and how much time and materials it will take.

You need a set of working plans with enough detail that a contractor could use them to remodel your bathroom while you were on vacation—a novel idea. The contractor shouldn't have to call you up every few minutes on a Cancun beach and ask where you want the shower or what color the walls should be. The plans need to be *that* complete and accurate. As mentioned in Chapter 5, the plans don't have to be professionally drawn—though it's not a bad idea—they just need to be a working plan for whomever remodels the room. You'll also need them for any required building permits.

Listing Materials

There are two reasons for accurate plans: to guide the remodeler and to guide the materials supplier. You can't precisely calculate costs unless you know what materials you need. For example, the price of a bathtub can range by hundreds or even thousands of dollars depending on the model. In addition, some fixtures take less time to install than others. A tile shower wall is more expensive and takes more time to install than one of plastic.

Fortunately, you can get help developing and pricing a materials list. Larger building material suppliers have on-staff estimators who can turn your detailed plans into a full materials list with prices. Some are more accurate and more helpful than others so you will want more than one materials estimate.

In addition, many popular home design and remodeling software programs have a component that develops a full materials list from your finished plans. Some help you get estimated local costs and even specify component model numbers to get the most accurate cost projection. What's amazing is that these programs can instantaneously give you revised estimates based on small changes you make to the design; you can find ways of saving money with various scenarios.

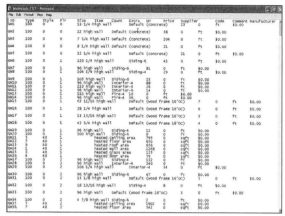

If you have a computer, developing your materials list in a spreadsheet program can save time and be useful to the estimator.

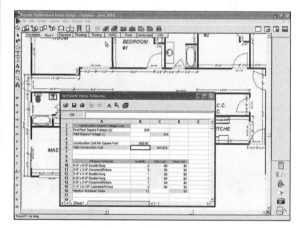

Many popular design software programs include a materials estimator function.

Calculating Labor

Labor costs may or may not be easier to calculate than materials costs. If you are providing all the labor you won't have to add in any labor costs. However, you probably should include it so you know the final value of your remodel. If you are hiring a remodeling contractor or doing some of the work and hiring subcontractors for specialized work you will want to estimate labor costs.

Of course, contractors can easily estimate how much time it will take to install a bathtub/shower module: 4 hours. How do they know this? Experience, yes, but there is another resource available to them—and to *you*! R.S. Means publishes books of estimated construction costs and labor time for most conceivable remodeling jobs. Other publishers have similar estimate books, but R.S. Means is the largest. They may not be easy for consumers to decipher, but they are an excellent resource for calculating labor and other remodeling costs. Copies are available through larger bookstores or online.

Waste Lines

You should estimate the *value* of labor, even if it is donated (friend, relative, co-worker, traded). Unless those donating are otherwise full-time remodeling professionals, estimate the value of donated labor at *half* to *one third* the value of professional labor. It's not that the results are less valuable, it's that it will take them two to three times as long to do the job.

In addition, I'll include some guidelines to help you estimate time and materials for projects described in this book. You'll find useful estimates throughout Part 2 on remodeling.

Considering Typical Budgets

As you estimate time and materials costs for a specific bathroom remodeling job it's nice to have a reality check, a range of costs for typical bathroom remodeling projects.

The cost depends on many factors, the most important being how extensive is the remodel. Is it a facelift, a makeover, or an addition? Is it a half, full, or extended bath? Let's take a look at some typical remodeling budgets for various bathrooms.

Half Bath Budgets

A half bath makeover includes complete replacement of the basic plumbing fixtures, flooring, vanity cabinet, and accessories including towel

bars. It also includes repainting the room. The estimated cost of a 24 sq. ft. half bath makeover as described is about $3,000. The estimate doesn't include tearing into walls to move major fixtures, replacing damaged subflooring, or making other major repairs that can add $1,000 or more to the final total.

If that's not enough money, you can add 50 percent or more to the total for a deluxe half bath makeover to include a single-piece toilet, ceramic tile floor, wainscoting (a design, typically of wood, that covers the bottom half of a wall), and upgraded vanity and fixtures. The total costs can be $5,000. Add more for any needed structural or plumbing repairs.

These calculations are based on typical labor markets, such as the Midwest, and not prices you'd pay for laborers in major cities or booming regions. In all of my examples, the cost of labor is about half the total cost. If you do it all yourself the materials cost is about $1,500 for the basic half bath makeover example and $2,500 to $3,000 for the deluxe version.

Waste Lines

As you estimate costs remember to consider the local labor market. In booming areas where housing costs are at a premium and the labor market is tight you will probably need to pay a premium for good labor and even some materials. In slower areas where labor is less expensive you might save as much as 20 percent on any labor costs, though probably not much on materials. If in doubt, ask a building contractor about the local labor market.

The reality check is that remodeling a half bath with a makeover will cost *about* $125 to $200 per sq. ft. depending on how fancy you want it. Compare that figure to your estimate.

Full Bath Budgets

Remodeling a full bath by replacing the toilet and tub, replacing flooring, painting the walls and ceiling, replacing the vanity, and upgrading the accessories will cost more like $6,000 if hired and about half that if you do all the work yourself. This estimate is for a standard 56 sq. ft. (7 ft. × 8 ft.) full bath. For a deluxe makeover with upgraded fixtures all around, plan on spending about 50 percent more, or about $9,000.

Full bath makeovers can actually be less expensive than half baths *per square foot*. Costs typically are less than double that of a half bath while the square footage is usually more than double. In the examples, the per-square-foot rates are about $110 to $160.

Extended Bath Budgets

An extended bath is one that has more amenities so the price goes up as fast and as far as you want it to. As a guideline, a standard makeover for a basic extended or master bathroom will be about $8,000 with the deluxe version coming in at around $12,000. If you are upgrading to a whirlpool, sauna, or other expensive amenity, tack on the extra cost. (Note: a basic bathtub is about $500 and a deluxe version is around $1,000, deductible from whatever your fancy-smancy tub will cost.)

The square-foot cost of extended bathroom makeovers is $100 to $150 or more in our examples. Again, cut the costs about in half if you're doing all the labor yourself.

Adding and Subtracting

These are guidelines for typical bathroom makeovers. Remember that facelifts—replacing surfaces and smaller fixtures—is much less expensive both in materials and labor. In fact, a facelift can cost 25 to 50 percent less than a makeover, and even less if you do all the work yourself.

On the other hand, if you need to add a room on to your home for a new bathroom you'll also need to add on the cost of new construction. That's more difficult to estimate, but you could start with a figure of about $200 per sq. ft. in addition to the makeover. The range is wide, from less than $100 per sq. ft. to add a dormer to over $300 per sq. ft. to install a small room with a new foundation and a revised roof system. Your figures will vary.

Making Your Own Budget

Now you have a realistic approximation of what your remodeled bathroom will cost. For example, if you're planning a makeover of a full bathroom of 8 ft. × 10 ft. (80 sq. ft.) you can start with a price of $110 per sq. ft., or $8,800. If you want to slightly upgrade the remodel, but not as far as deluxe, you can add 20 percent and come up with $10,560. If you've opted to buy all materials and install them yourself you can cut that number by at least half for an approximation of no more than $5,280. If you ultimately decide to hire an electrician for the new fixtures you can add that to the total.

Of course, an approximation isn't an estimate. It's a ballpark figure, a reality check. If you get a bid from a contractor (discussed in Chapter 8) that's 50 percent more than your approximation you know something's wrong. Either the contractor's fees are too high, there are design issues that make it a tougher job than it looks, or the material or labor is overpriced.

How can you make sure that your estimates and bids are accurate? You can clearly describe the project, review the plans for potential high-cost problems, accurately calculate materials, and estimate labor.

Describing the Project

The building plans you developed in Chapter 5 should be sufficiently accurate to guide a professional builder. That means all components and fixtures are clearly identified and specified (described) in a way that makes sense to builders. For example, windows should be described by size (3 ft. × 4 ft.), style (double-hung, casement, etc.), type (double-glazed, triple-glazed), trim (vinyl clad, metal clad), and other components (screens, drip cap). The cost, the installation labor, and the results all depend on your accurate description of what you want in your remodeled bathroom.

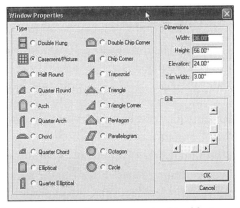

If using design software you may be able to easily specify the type and size of window for the plan.

> ### Potty Training
> Specifics make planning the job easier. Be as specific as possible as you describe, design, and plan your remodeled bathroom. Approximating the location of a lavatory, for example, can become costly as you later have to move plumbing to match the cabinet you decided to purchase.

In most cases all a builder needs is a set of single-dimensional plans that indicate locations and dimension of fixtures and materials. In some remodels there will be details that can't be communicated on a flat plan. That's when an architect develops detail drawings to guide construction. A detail typically is a cross-sectional drawing that shows what the component looks like if cut in half. For example, an alcove in a shower wall may best be described by a detail drawing to show the framing, the plumbing, and subsequent tile and fixtures.

There's no building law that says you can't describe things in words if they clarify and don't contradict the drawn plans. So, if you're handy with words, attempt to describe your vision for a specific feature, making sure that it offers clarity.

Reviewing Plans

No matter who draws up your bathroom remodeling plans, it's a good idea to have them reviewed by someone else. If you draw them, hire an architect or a contractor to look them over before starting the work. They may see some things that you totally missed or weren't clear about. If you've hired an architect to do the work, have a contractor check the plans. If you are doing your own work, have the architect explain the plans thoroughly and answer any questions you have on the project. Get as much detailed information as you can early in the process. If you call the architect during the weekend when you're remodeling, he probably won't answer the telephone.

If you're a true do-it-yourselfer you can opt to take your plans to the local building department for a preliminary plan check. Make sure to schedule an appointment. An inspector *might* be available to discuss the plans with you and give you some tips on meeting local building codes. Remember, however, that building codes cover health and safety issues. They aren't the design police.

The point here is to make sure that your remodeling and construction plans meet local building codes and don't have design issues that can dramatically increase costs or reduce your contentment with the results. Make sure that plans are accurate and clear.

Calculating Materials

Accurate plans lead to accurate cost estimates. In addition, clear plans mean that whoever estimates material costs for you doesn't have to guess what you want. If you specify a toilet by model and color, for example, not only can the estimator be more accurate but you can easily compare bids between various material suppliers.

If you're opting to use a contractor for your remodeling project you can help by reviewing all materials bids received to make sure they comply with the materials list. It's a double-check that can save hundreds of dollars as well as give you greater participation in your home.

Heads Up!

Even if you're just remodeling a small bathroom yourself, keep a first-aid kit handy. It takes only a moment to injure yourself; make sure it takes little time to take care of it. The kit can include basic first aid supplies and cost less than $20, available at building material or auto parts stores, or put one together yourself.

Can you change your mind? Yes, for a price. If you change your mind on a remodeling component before it is ordered the plan change may be minimal. If you wait until the new tub is delivered to decide that you want another color the cost in time and money will be much higher. If the change makes a difference to the cost or remodel schedule a contractor probably will ask you to sign a change order document to get any changes you make in writing with your signature. You will be billed for any changes that impact the contractor's time or material costs.

Estimating Labor

As mentioned, many professional contractors use experience or time books to help them estimate the time required to perform most remodeling tasks. For example, if they have not installed a specific model of whirlpool before, a time book or their wholesaler may tell them that it takes 3.5 hours for the typical installation. They then multiply their hourly rate by 3.5 to arrive at an estimate of labor costs for the installation.

How much time will it probably take for you, a do-it-yourselfer, to install the same fixture? Unless you have prior remodeling experience with plumbing fixtures, double the time it would take a pro, 7 hours in this example. Why? Because you're going to have to read the installation instructions first, a task most pros don't do, and you're going to work slower and more carefully.

Finalizing the Budget

You'll soon write a check—or ask a lender to write a check—for the cost of your bathroom remodel. No matter who actually writes it, the money is yours. Make sure that the remodeling plan documents are as accurate and complete as possible. Take a final review of your goals, your design, and your options.

If you're using a lender (see Chapter 7), you or a contractor will develop the final cost figures and submit them to the lender. Even if *you* are the contractor *and* the lender, it's a good idea to produce a written budget that goes with your remodel plan. Think of it as your remodeling map.

Contractor or DIY?

Have you decided yet whether you will use a contractor or do the remodeling yourself? The answer greatly impacts the final cost. As covered earlier in this chapter, the cost of labor is at least half the total cost of a typical remodel. If the needed bathroom remodel costs just won't fit your current finances, consider doing some or all of the work yourself to save money. You'll also learn new things, which can be fun.

I'll show you how to find and hire a qualified bathroom remodeling contractor in Chapter 8. For now you need to answer the question: Do I need to or want to hire a contractor? The answer dramatically changes the total cost.

Plan A or Plan B?

Most people who remodel have a Plan A and a Plan B. That is, if they can get sufficient financing they will upgrade the new tub to a whirlpool; otherwise they will replace the existing tub with a free-standing unit. Or the options may be to upgrade the toilet if there is no need to make floor repairs once remodeling has begun.

Options are good. Make sure that your plans, estimates, and budget match the alternate plans. Also make sure that the remodeling timeline allows flexibility between the various options so you can "change your mind" without ripping out something you just installed.

Cutting Costs

If the estimated cost is still over what you can afford to pay for the bathroom remodel, consider additional ways to cut costs. For example, if plans call for moving a doorway over a foot—a relatively difficult and expensive project—consider not doing so. Look for options.

You also can cut costs by shopping wisely. If you have the time and space you can wait until your favorite building material supplier has a clearance sale to buy major fixtures. Or you can find another supplier with a better price even if it includes a trip to pick up materials.

Another cost-cutting option, of course, is to do more of the labor yourself. That may mean helping the flooring installer or installing your own faucets or even installing the tub yourself though you hadn't planned to do so. As long as you are a help rather than a hindrance you may be able to save some money.

The worst cost-cutting option is to buy cheap materials or hire cheap labor. You may get lucky and it will all work fine at a lower cost. However, most professionals agree that "cheap" is false economy.

> **Waste Lines**
>
> On a tight remodeling budget? Find out if there is a salvage center in your area (check the telephone book). Salvage centers typically have good quality fixtures and building materials at lower costs than new. You may even find an antique tub, shower, toilet, or fixture that matches your new bathroom. It's the ultimate in recycling.

The Bottom Line

You've done your homework. You know how much your bathroom remodeling project will cost and you've carefully reviewed your options. Congratulations! All you need now is the money. That's what we'll cover in the next chapter.

Meantime, remember that this project is a focused attempt at improving your life. Its purpose is to give yourself and living partners more amenities, more space, or more privacy. So enjoy the process. You'll definitely enjoy the results.

The Least You Need to Know

- Make sure that you have detailed plans of your remodel project to ensure an accurate cost and time estimate.

- Verify your remodel budget against a typical per-square-foot estimate to make sure they are realistic.

- Remember that the largest cost of a remodel is labor and decide how much of the job you will do yourself.

- Time spent on managing your bathroom remodeling budget can be valuable to the end result.

In This Chapter

- ◆ Remodeling as an investment
- ◆ Finalizing bathroom remodeling costs
- ◆ Finding the money
- ◆ The inconvenience factor: Can you live without the bathroom for a while?

  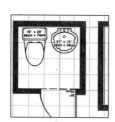   

Chapter 7

Getting the Money

Money is the root of all remodeling. Even if you're doing all the installation you still need money for new plumbing and electrical materials. You still need *some* money.

Where are you going to get the cash? If you've already saved up for this job, congratulations! Or maybe you hit a smaller lottery jackpot and, after taxes, you have just enough to remodel the bathroom. Or maybe you're not sure where the money is coming from.

Even if you already *do* have a source of bathroom remodeling funds, read through this chapter. It offers something just as valuable as money: ideas. You'll see how thousands of other successful remodelers have cut costs and increased the value of their home with a new bathroom—and how they paid for it.

Is It Really Worth the Trouble?

Bathroom remodeling typically isn't something folks put down on their list of Favorite Vacation Pastimes. It's work. Even if you're hiring someone to do the work while you run off to relax in the sun you'll still need to plan and pay for the project. So it's a fair question to ask yourself: Is a bathroom remodel *really* worth the trouble?

The answer depends on whether the cost is an investment in your home. Will you live in the home long enough for the equity increase to overcome the cost? Is the cost an investment or an expense? Most important, keep in mind the long-term benefits you expect your remodeled bathroom to bring to you and your family.

How Long Will You Live in Your Home?

It's impossible for most folks to know exactly how long they will live in a specific home before moving on. Jobs change. Life makes unplanned announcements. Other investments ripen or go sour, altering finances. However, what's your best current estimation of how long you expect to live in your current home?

The answer is important because bathroom remodeling can be an expensive project, typically costing $3,000 to $20,000 or more, as you've seen in the previous chapter. In addition, you may not get all of your investment back if you must sell your home with a partially remodeled bathroom. In areas where home values appreciate fast, remodeling a bathroom can be an excellent investment that multiplies potential profits from the sale of your home. In places where home values increase slowly—or actually go down—some of the money spent on remodeling a bathroom may be lost until the home appreciates more in value.

Is It an Expense or an Investment?

Your answer to this question depends on what you're doing with the bathroom. If you're upgrading your home's services, it's an expense and no financial payback is expected. However, if you're remodeling your home with an office and adjacent bathroom for your home-based business, it's an investment. That is, the expenses might be written off your income tax obligation or qualify you for a higher mortgage amount because of the income that it will bring in. If you're getting into areas of taxation, spend a few bucks to talk with a tax accountant with small business or real estate investment experience.

Waste Lines

Upgrading a bathroom for access by the disabled may qualify as a medical expense, deductible from your income tax. Discuss your options and costs with a tax advisor.

Are the benefits worth the cost? Typically, yes. And by planning smart and participating in the bathroom's remodel and construction you can ensure that you will benefit from increased livability and increased value.

How *Much* Money?

Chapter 6 showed you how to budget the job and get estimates for the plans you earlier developed. Let's pull it all together now into a final-final number that you can literally take to the bank.

Getting a Bid

Even if you're planning to remodel the bathroom yourself and you have a source for funding, consider asking one or two contractors to bid on the job. If nothing else you'll learn how contractors work. They may bring up some options or some problems you may not have considered.

How can you get a bid? Many consumers start with the telephone book or at a local building material supplier's store. Check ads in the phone book and look for business cards on the store's bulletin boards. You're looking for licensed contractors who have experience remodeling bathrooms.

The next step is to simply call up the contractors and ask for a bid. If they try to give you a bid over the telephone, don't bother with them. A bid should be in writing *and* be based on a visit to your home to see what they'll be dealing with.

Get at least two bids on your remodel job. I'll offer additional advice on hiring contractors and experts in Chapter 8.

Making Your Own Estimate

You have an estimated budget of the costs of your bathroom remodel but it may not be very formal. Type or neatly print an estimate of all materials and labor costs. Include actual remodeling plans. This is your formal summary of the design and costs of your bathroom remodel.

You can include additional notes in your remodel job file, including fixture specifications and a list of suppliers. Also include "before" photos and your bathroom remodel notebook. This is your working file.

If your remodeling job is extensive also include an estimate of the value your home will increase with the new bathroom. You can do your own homework, analyzing other homes in the area, or you can ask a real estate agent or hire an appraiser to give you their estimate of how much the home will increase in value with your remodel. Remember that you don't pay an agent for this advice so it is an educated guess, but an appraiser is paid so it is a professional estimate and more accurate.

Finalizing Your Budget

Finally, make extra copies of the contractor bids you've accepted or the cost estimates you've developed. You'll need copies if you are working with any lender that uses your home as equity for the loan.

That's it. You're ready to consider the next big question: How am I going to pay for this bathroom remodel? Maybe you already have a funding source, but keep reading as you may discover more options that can save you money.

Paying for Your Bathroom Remodel

Where are you going to get the money? If you don't have a rich Aunt Helen who will write you a check for the bathroom remodel—and even if you do—there are many options for paying for the job. They include cash and various loans.

Financing is a part of everyday life. Though debt can quickly turn to servitude, using other people's money can be a smart move. Borrow wisely.

Cash

Cash is good. In fact, if you can pay for your bathroom remodeling project in cash you typically can earn a small discount from most suppliers. The discount can range from 2 to 10 percent of the total. To get the discount, speak with the manager of various material suppliers in your area to find out who has the best *net* (after discounts) prices for cash transactions.

Some folks with smaller bathroom remodeling projects use cash on a pay-as-you-go basis. That is, they set aside a specific amount each month from savings, regular income, or a combination of both, and buy what they need once a month. They then install it over the coming month. The remodel is then paid for by the time the work is done. No interest.

Consumer Loans

You can get a consumer loan for what you need either through a consumer loan company or on your credit card. The problem here is the higher interest rate. If you don't pay the loan off they come after you, not your house and new bathroom. From the lender's standpoint there is less security for the loan, your promise to pay. That's why they get the higher interest.

However, for smaller remodeling projects that you plan to pay off in just a few months a consumer loan or credit card purchase is typically easier to get than refinancing your house.

Mortgages

A *mortgage* is a loan that uses real estate as security. If you don't pay they come and take your house. Because the security typically is worth more than the loan amount, the lender's risk is reduced. That means the interest rate charged is lower than with higher-risk consumer loans.

>
>
> **Bathroom Words**
>
> A **mortgage** is agreement between a lender and a buyer using real property as security for the loan.

The primary mortgage against your home, the one you probably signed when you bought your house, is called the first mortgage. If you want to use some of the equity in your home as collateral for your bathroom remodel you can refinance your first mortgage *or* get a second mortgage.

What's the difference? You'll be paying off both mortgages simultaneously. It's just that if you default on your mortgages, the lender that holds your *first mortgage* gets paid off first. So interest rates on a *second mortgage* may be slightly higher than on the first because of the lender's higher risk.

The good news is that the increased value of your home with a newly remodeled bathroom typically can be the equity you need to fund it. Should you refinance your first mortgage or get a second mortgage? Good question! The answer depends on the equity you now have in your home, the value of your remodel, how much you need to get the job done, and current mortgage rates. You can start your search for the best answer by talking with your current mortgage lender and your primary bank.

Inspections and Draws

If your remodel project is small and you are self-financed (cash, consumer loan, or credit card) you won't have a mortgage. With a mortgage often comes lender requirements such as inspections. If, for example, you have a $20,000 second mortgage to finance remodeling your bathroom, the lender probably will define specific points at which money will be paid to you, a supplier, and/or a contractor. The lender probably won't put all $20,000 in your checking account and let you spend it on a new car. Its purpose is to increase the value or equity you have in your home.

> **Potty Training**
>
> Need to find the best contractor in town? You can't! He's too busy to talk with you. However, busy contractors often will recommend other contractors that they know are qualified but not as busy. So start at the top when looking for a bathroom remodeling contractor.

Inspections may be set for when all walls are framed, then once the plumbing and electrical systems are installed, and finally once the bathroom is finished and habitable. The lender may release chunks of money at each step, called *draw requests*. The challenge for you and/or your contractor is to make sure the draws cover the expenses to that point. If the plumbing draw is $4,000 and the plumbing contractor's bill is $6,000, for example, someone (guess who) will need to come up with $2,000 to pay the contractor. If you're working with a general contractor who is managing the project, he or she will coordinate the work and the draws. If you're doing it yourself you'll need to coordinate them.

Bathroom Words

A **draw request** is a monthly request by a contractor or do-it-yourself homeowner to be paid for the materials and labor installed into the project during the previous 30 days, to be drawn from the construction loan.

About Liens

A lien is legal charge against a project. A mechanic's lien is one made by a contractor and a materials lien is made by someone who supplies materials for the job. Liens tell everyone "I have done some of the work or supplied some of the material for this job and I won't release my rights to get paid until I actually *do* get paid."

Lenders who make payments to contractors and suppliers will require that they simultaneously sign off on the lien saying that it has been satisfied and that they have no additional rights to the property. If you are self-financing the bathroom remodel, make sure you get the lien released when you pay a contractor or supplier. Most contractors, suppliers, and stationery stores have the forms.

If you'd like to know more about mortgages, lenders, liens, and other financial aspects of homes, get my book, *The Complete Idiot's Guide to Building Your Own Home* (Alpha Books, 2002).

Life in a Construction Zone

I've covered how much remodeling your bathroom will cost in financial terms. Now let's consider how much it might cost in living terms.

Here are three questions, and discussions of each, that you should consider before you actually begin the remodeling process. I don't want to encourage projects that break up otherwise happy homes!

How Inconvenient?

Face it: Remodeling your home's bathroom will be inconvenient. Maybe you're doing the work in the evenings and on weekends. Or you've hired a contractor to do the work during the week. The side yard will be torn up by a bulldozer clawing for access to your foundation. Or it will be stacked high with lumber, plumbing fixtures, and other expensive things.

Potty Training

Remodelers recommend that you document your project in photographs. Not only does it illustrate progress, but it also helps you find problems later. If a leak develops in the ceiling or walls you can pull out the progress pictures and determine what might be the source—and how best to get to it.

I don't know your exact situation, what you'll be doing in your bathroom, who is going to do the work, or how you'll pay for it, so I can't guess how inconvenient the job might become. But I can make an educated guess! Unless you are hiring an angelic contractor who will finish the job while you're off to the Bahamas, chances are you *will* face inconvenience. The driveway will be blocked. The plumbing will be turned off for a few days. Needed materials will be back-ordered, putting progress on hold.

Consider your project from this level: inconvenience. Think of it as not only your personal inconvenience but that of other members of your living group. Then, do what you can to minimize the inconvenience. You may decide to hire someone to help you get the job done faster. Or you may want to take a short vacation while a contractor tears up your driveway. Do whatever you can to minimize the inconvenience of remodeling to you and those with whom you share living space.

How *Long?*

The famous "Murphy" of "Murphy's Law" fame must have realized his truth while attempting to remodel a bathroom. "Anything that can go wrong will, at the worst time." Plan as you might, a moisture problem, the wrong materials, a late inspection, an unexpected cost increase, or some other life event can s-l-o-w progress on remodeling your bathroom.

The best advice is to do the project as quickly as possible, even if you have to do it in stages. Delays become inertia and projects don't seem to get done. Life intercedes and that bathroom project is moved to the back burner. The solution is to make sure you have not only the money, but the available time to complete your bathroom.

Heads Up!

Living in a construction zone can be hazardous to your health—and to the health of kids, pets, and other critters. Make sure that the last job of each day—yours or your contractor's—is to make sure that the site is safe. Otherwise, lock the door to the bathroom.

What If?

Stuff happens. What if you just cannot finish your bathroom remodeling? The money runs out. Illness stops the labor force. A new job opportunity requires that you be elsewhere—next Tuesday! What can you do?

The answer is: the best you can. If you've broken your project into subprojects you'll always be near a finishing point. If not, consider hiring (or begging) someone to continue the work for you. If all else fails, just stop where you are. You are remodeling your bathroom to increase enjoyment for you and your living group. If the task becomes counterproductive, take some time to rethink the project and how it fits in your life. Maybe you just need a short vacation from it.

Gosh, I hope you don't think I'm trying to talk you out of remodeling your bathroom. I'm really not! I want you to consider both the plusses and the minuses of it *before* you actually start buying things and installing them. Make sure you've planned thoroughly, estimated costs accurately, figured out exactly how you're going to pay for it, and considered potential problems. Doing so will increase your chances of success as well as your enjoyment of your new living space. And that's what it's all about.

The Least You Need to Know

◆ Consider the financial and living implications of your bathroom remodeling project.

◆ Get bids from remodeling contractors even if you plan to do all the work yourself.

◆ Small remodels can be paid by cash or consumer loans; larger remodeling projects may require a new mortgage on the home.

◆ Consider the inconveniences of being without the bathroom for the duration of the remodel.

In This Chapter

- ◆ Deciding how much help you really need
- ◆ Finding a designer or architect for your remodel
- ◆ How to choose the right contractor or subcontractors
- ◆ Getting the best price and delivery on remodeling materials

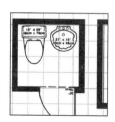

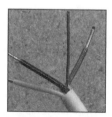

Chapter 8

Hiring Experts

The key to success on any new project is getting good advice. This book offers an inside look at how bathrooms are remodeled as well as what to consider when you do so. But this book can't actually *build* your new bathroom. You or someone else is going to have to do the actual work. Aw!

So who do you need to hire as help? Maybe nobody. Depending on your remodel job, your comfort level with tools, and your budget, you may or may not need to hire a bathroom designer or architect, remodeling contractor, electrical or other subcontractors, or laborers. You probably will have to select a materials supplier. In any event, you need to know what each expert does so you can decide whether it's an expertise you need to hire.

How Much Help Do You Need?

Want to save some money? As much as *half* the cost of your bathroom remodel? Then do some or all of the work yourself! It's really not that difficult. I'll show you how in Part 2 of this book. Meantime, take a reality test by answering the following questions about your bathroom remodeling job:

- ◆ Do you have a plan for your bathroom remodel including design and costs?
- ◆ Do you have the time and energy to complete this remodeling project?
- ◆ What other do-it-yourself projects have you completed and what did you learn from them?
- ◆ Is this project a facelift, makeover, or addition?
- ◆ How much of this remodel are you comfortable doing yourself?

◆ Do you have the funds to hire a remodeling contractor or does your budget require that you do some of the work yourself?

◆ Do you have friends or relatives with remodeling experience who can advise you on difficult areas?

◆ Do you have a team of living partners or friends who can provide additional hands when needed?

◆ How strict are local building codes (see Chapter 9)?

◆ Is this job big enough to earn a discount from a local materials supplier?

◆ How motivated are you to do this remodel yourself versus hiring some or all of it done for you?

From the answers to these and related questions you can determine your needs, resources, and preferences for doing it yourself or hiring it done. Consider that no matter what you decide *someone* has to do the work, either you or an expert. So learning more about what each expert does will help you make an informed decision on whether to hire him or her—or yourself.

Hiring a Bathroom Designer or Architect

Are you considering hiring an expert to help you design your bathroom remodel? Architects and designers offer a wide variety of services. They can do everything from critiquing your own bathroom remodel design to acting as the construction manager. Many specialize in aspects of construction and remodeling such as bathrooms or kitchens.

Where can you find a qualified remodeling designer or architect? Start by looking in the local telephone book. Also ask friends, bathroom material suppliers, lenders, and contractors you've met. You don't want to find out that you hired the fourth-best architect!

How can you hire the best architect to design *your* new bathroom? Here are some questions to ask designers and architects you consider hiring:

◆ Does your firm have a valid state architect's license?

◆ What national and local design or architectural groups are you affiliated with?

◆ How long have you been in business?

◆ What bathroom remodeling jobs have you completed recently?

◆ Do you work exclusively with contractors or do you have experience designing for homeowners directly?

◆ What are some of the challenges you've overcome in remodeling bathrooms for clients?

◆ Can you furnish me with plan examples and references?

◆ How well have your remodeling estimates matched actual building costs?

◆ What services do you provide during the design, bidding, and construction phases?

◆ What are your fees for each of these services?

◆ How do you calculate your fees?

◆ Who in your firm actually does the work and who oversees?

◆ Can you recommend specific contractors, subcontractors, suppliers, and lenders for my remodeling job?

◆ What additional services and fees will my project require?

◆ How will design changes be made and charged?

You get the idea. Keep reading to learn more about what specific experts do and determine whether you need them or not.

Potty Training —————
Don't forget your friends! When searching for a qualified bathroom remodeling contractor, ask friends, neighbors, and co-workers if they have had experience—good or bad—with local remodeling contractors. Can they recommend anyone?

Hiring a Contractor

For many remodeling jobs, hiring the contractor is the most important step. A qualified and efficient remodeling contractor can save you headaches and more than earn their fee.

Remodeling contractors are a specialized breed. Let's take a look at what they do before deciding whether and how to hire one.

The Remodeling Contractor's Job

Most remodeling contractors are actually general contractors. That is, they have extensive construction *and* business experience in many aspects of construction and remodeling. They could build you a house, remodel it, repair it, or raze it. They've probably passed state examinations and are licensed.

Some bathroom remodeling contractors are actually specialists who have expanded their services. For example, many bathroom contractors started out as plumbing contractors and have morphed into bathroom specialists to meet the needs of customers.

So what does a remodeling contractor do for his or her money? Of course, that depends on what *you* hire the contractor to do. What's included in the contractor agreement? In most cases, your remodeling contractor will:

◆ Supervise all aspects of the work done at the remodeling site

◆ Hire, supervise, pay, and fire subcontractors as needed to get the job done

◆ Coordinate getting building permits and any variances

◆ Buy all materials and supplies needed in construction

◆ Make sure that the site is inspected and approved by the building department

◆ Make sure that all subcontractors are legal and have the needed insurance (like worker's compensation insurance)

For this effort, the contractor typically gets 10 to 25 percent of the total value of the project. The contractor subtotals the costs of materials and subcontractors, then adds the management fee. Alternately, the remodeling contractor can be hired to manage the project for a specified hourly rate. Or a lump sum may be agreed upon.

Finding the Best Contractor

How are you going to find the *best* contractor you can afford to remodel your bathroom? Referrals are the first place to seek remodeling contractors. Ask people you know who have had bathrooms remodeled in the area recently. Talk with local building material suppliers. Ask your lender and any subcontractors you know for recommendations.

How do you interview a remodeling contractor? Face to face if possible. Lunch, long coffee break, or an office or job-site appointment are good meeting grounds. Of those, seeing a contractor at a job site can be enlightening as long as your presence isn't distracting. What should you ask candidates?

◆ What can you tell me about your license and remodeling experience?

◆ What type of contracting license do you have and what is the license number?

◆ What experience do you have with remodeling bathrooms?

◆ What references can you give me?

◆ Can you post a *performance bond*? If not, why?

Bathroom Words

A **performance bond** is like an insurance policy that guarantees to the owner that the contractor will performed in accordance with the signed remodeling contract. The contractor buys the bond in favor of the owner.

◆ If you hire employees, may I see your worker's comp policy?

◆ How do you go about hiring subcontractors?

◆ How do you schedule a bathroom remodeling project?

◆ Do you typically meet your completion schedules?

◆ What lenders do you prefer to work with? Why?

◆ What suppliers do you prefer to work with? Why?

◆ What things do you need to make a bid on this project?

◆ Can I have my attorney look over your standard remodeling contract?

◆ If I hire you, how can I help make sure the project goes smoothly?

Remember, you want to ask open-ended questions that aren't answered by "yes" or "no." You want to hear explanations and find out how well the person communicates. Of course, there are many more questions you could ask depending on your bathroom plans and how you will participate in the remodeling process.

The contractor's license number can come in handy. States that license contractors typically keep a file you can access if you know the contractor's license number. You can do without it, but it makes the job much easier if you know the number. The file will have information about when and where the license was issued as well as any complaints or commendations filed.

Calling for Bids

You've been busy. You've found and interviewed some remodeling contractor candidates and selected two or three you think you can work well with. It's time to talk money!

The key to accurate bids is accurate specifications. That is, contractors don't have to guess when working with clear remodeling plans. They know how much the materials will cost, how many hours it will take for each stage, and how long until it is done. Contractors who have to guess will always lean toward the higher price.

So your first job is to make sure that each remodeling contractor has the same complete set of prints and specs. The cost of extra prints is small compared to the total cost of the project, so if you really want an accurate bid from a contractor, make sure he or she has all the prints and specs needed.

Be sure you know what's included in the bids. Building permit fees? Travel expenses? Site security? Cleanup charges?

Bathroom Words

As you read this chapter and come upon any unfamiliar construction terms, be sure to look them up in Appendix A.

Also ask for a start and finish date in the schedule. It will make a difference as you pay interest on a remodeling loan or need to schedule the rest of your life around this project.

Get job responsibilities in writing on the bid. Who will call for inspections? Who notifies the lender to release a draw? How will changes be handled? These questions must be answered in the building contract, so ask the contractor to include them in the bid.

You can narrow the contractors to one or two of the best bids and ask for a rebid based on minor changes. You can also tell contractors what you're looking for in the final bid. "How can I get a final bid that's 10 percent less without cutting quality?"

Be careful; don't get cheap. If you've developed a working relationship with these bidders, you have an idea what it will take to get a fair bid. And that's what you want.

Waste Lines

You'll hear contractors refer to *substantial completion*. It's the time at which the contractor feels that he or she has essentially finished the project, but before the final inspection. Make sure your contract includes in writing what is to be done before final payment for services.

What's Included in the Contract

Did you find a qualified remodeling contractor to take on your bathroom? Great! Let's get it in writing.

A typical remodeling agreement will include some important specifics. Attached to it (by reference) will be the remodeling plans and probably the final bid. The agreement itself will include:

- ◆ Contract date and parties
- ◆ Starting and completion dates
- ◆ Conditions (who does what, who pays for what, who approves what, how changes are made and paid for, and so on)
- ◆ Contract sum
- ◆ Progress payments (matching the lender's draw schedule, if any)
- ◆ Terms of final payment
- ◆ Contract termination terms
- ◆ Anything else agreed upon that should be in writing

Don't sign anything under pressure. If you're not comfortable with the price, terms, and conditions, stop and think it through. Will this contract take you to your goal? What's the worst that can happen? Is it covered by the contract? If it will make you more comfortable, have a contract law or real estate attorney review the building agreement prior to and during final signing.

Hiring Specialty Contractors and Laborers

A general or remodeling contractor typically is a boss who may hire other contractors—called specialty contractors—to do much or even all of the work. A contractor also may hire laborers to help out with the low-skill work. If you are your own contractor on this remodeling job you need to know how to hire specialty contractors and laborers to make the job go easier. You don't need more problems.

There is a wide variety of specialty contractors from which to choose, each with an area of expertise. Let's take a look at those specialties and the contractors who do them.

Heads Up!

Many states and provinces have worker's compensation programs, paying workers if they are injured. The compensation typically is too small to encourage widespread abuse, but some workers abuse it anyway. Note that if you hire "employees" you may be subject to worker's compensation laws and be required to pay toward the insurance. The construction industry is one of the most dangerous and has higher worker's comp premiums than others. Before hiring anyone contact your state labor board through the number in the government section of your regional telephone book.

Foundation Contractor

The foundation contractor is responsible for preparing the site for and installing the foundation by the approved building plans. The foundation contractor might work with the grading contractor to dig for footings, or might have his or her own crew do it.

The foundation contractor will install forms, install reinforcement bar (rebar), work with plumbers or other contractors who need to bury their utilities in the concrete, and get everything for the concrete. The foundation contractor may hire a concrete contractor to pour and finish the concrete, or he or she might have his or her crew do it.

Framing Contractor

If you need a new wall or to have one moved, you can hire a framing contractor. They can install joists, floor decking, walls, rafters or roof trusses, and roof sheathing. The framing contractor may also install doors, windows, and exterior siding.

If you decide to hire a general contractor, he or she might do the framing and maybe even a few of the other specialized jobs. You don't

want to hire contractors outside of their expertise, but many are sufficiently knowledgeable of the entire building and remodeling process to do more than one job.

Plumbing Contractor

The plumbing system must be roughed-in or started early in the remodeling process. The plumbing contractor will install runs or horizontal pipes and risers or vertical pipes for the water, sewer, and drain systems.

If something will later be built over the top of these pipes (such as concrete slabs), the plumbing contractor will install primary plumbing as the foundation is being built. Otherwise, the plumbing contractor can rough-in plumbing later. Once the framing is done, the plumbing contractor will be back to install supply, drain, and vent pipes throughout the house. Finally, the plumbing contractor will return after the walls are closed to install plumbing fixtures and hardware.

Electrical Contractor

The electrical contractor starts at the electrical service panel, installing new wiring as needed for electricity to be delivered to the remodeled bathroom. From the panel, wires are run throughout the house through framed walls, floors, and ceilings. If the walls are closed, the electrician may opt to run wiring under the house or through an attic to reach the bathroom.

Finally, the electrical contractor returns when the walls are closed and installs outlets, switches, and cover plates. This occurs once the insulation and drywall are installed over the wall framing.

Mechanical Contractor

One or a group of contractors may tackle the heating, ventilation, and air conditioning (HVAC) systems. They design the systems and install ducts, furnaces, fans, vents, and compressors. Finally, they install filters, registers, and grills.

In larger communities you may find that the furnace is installed by one contractor and the ducting for central distribution is installed by another specialist. Radiant floor and ceiling heat systems have their own specialized contractors.

Drywall Contractor

Drywall is a sheet material made of gypsum plaster covered in thick paper, cut into sheets of 4 ft. × 8 ft. or 4 ft. × 12 ft. The sheets are nailed or screwed to wall framing, then the seams are taped so the walls look seamless. Some drywall then has a texture applied to the walls and/or ceiling.

Alternately, walls are plastered with a wet plaster installed over wood lathe or other surface. The plaster is hand-contoured, then dries. Plaster contractors are specialists and more difficult to find.

Potty Training

How can you find qualified drywall and wet plaster contractors? Drive by homes under construction or remodeling and look for truck signs that tell you who is working there. Also ask friends and material suppliers for their recommendations. Remember, though, that the better contractors may be quite busy and not able to accommodate your remodeling schedule.

Flooring Contractor

The flooring contractor will install carpet, vinyl tile, solid or laminate wood, or other flooring materials following the building plans. Flooring typically is attached to the subfloor with adhesive or nail strips, or it floats unattached using an interlocking system to keep it in place.

Some flooring contractors specialize in one or two types of flooring, such as carpeting, and leave the others to someone else. They are specialists among specialists.

Painting Contractor

Painting contractors apply paints and other finishes to surfaces. Application is with brushes or sprayers depending on how much surface is to be covered and what is nearby.

There are even specialists among painting contractors. Some do exteriors or interiors only. Others specialize in painting high places. Still others work with special finishes or painting techniques.

Selecting the Best Subs

As with hiring general contractors make sure the specialty contractors are qualified, licensed, and that you can work with them without problems. Many are notoriously overbooked and overworked. Because your bathroom remodel might be a smaller job than others and because the contractor will work for you once and work with other contractors many times, you might not get the specialist's full attention.

Review the suggestions for hiring contractors earlier in the chapter. Ask others for referrals. Follow up on references. Check out licenses. Don't sign anything until you're sure what it is.

Hiring Good Laborers

Depending on your own skill level, you may decide to take on more of the remodeling yourself and hire laborers to help. Where can you find such laborers?

◆ Newspaper classified service ads
◆ Local employment offices
◆ Employment services
◆ Qualified friends and relatives

- ◆ Moonlighting construction workers
- ◆ College employment offices (especially those with construction trade classes)
- ◆ Referrals from material suppliers

First, figure out what you need and write a description. For example, "Need experienced construction worker to help remodel a bathroom. Must be able to work safely." A clear description will help direct you to the best place to find such a person.

How much will you pay your laborer? That depends on the value of the work done and the local labor market. Ask your lender, contractors, material suppliers, employment offices, and labor halls what it will take to get qualified help. Then plan to pay 10 to 20 percent more for better help.

Hiring Materials Suppliers

You may need lots of materials to remodel your bathroom. The list can include plumbing fixtures, electrical fixtures, cabinets, lumber, drywall, fasteners, doors, windows, and lots more. Where are you going to get all this stuff when you need it and at the lowest price?

First, review your bathroom remodel notebook for names of material suppliers that you've discovered or heard about. Maybe a contractor recommended them or you've seen local ads that looked like their pricing is lower than others.

Then contact each supplier about bidding on your project. As with getting bids from contractors, suppliers will need specs and plans. Make sure they get a comprehensive materials list if available. Many architects and plan services will provide this list with remodel plans. Most building material suppliers can develop a comprehensive materials list from your plans. It used to be done the old-fashioned way with

paper and pencil. Today, a computer spits out pages of materials requirements from a plan. However, remember the rule of computers: Garbage in, garbage out. If the plans aren't complete or the data isn't entered in correctly, the materials list will be inaccurate and the bid will be incorrect.

Chief Architect and other residential design software programs can develop a materials list for remodeling projects.

Materials lists can be exported to a spreadsheet.

Before selecting your primary materials supplier, interview one of the salespeople in charge of remodeling accounts. You want to know about quality, pricing, discounts, delivery, terms, and advice:

- Are preferred brands (if any) readily available from the supplier? If not, can they be easily ordered?
- What are your prices on (list a few specific products such as a shower unit or counter top)?
- What discounts and terms are available to licensed contractors?
- Do you offer the same discounts and terms to do-it-yourselfers?
- What discounts are available to do-it-yourselfers for cash-and-carry?
- How soon can you typically deliver materials to my remodeling site?
- Will I have a specific salesperson assigned to my account (preferred)?
- Who would I talk to if I can't come to an agreement with a salesperson?

That covers it. Hiring the best contractors, subs, laborers, and material suppliers can make your bathroom remodeling job go much easier. As any businessperson knows, hiring good help is the key to success.

The Least You Need to Know

- Remodeling contractors are experts in construction as well as in business.
- You can directly hire specialist contractors to install plumbing, electrical, framing, flooring, and other components of your new bathroom.
- Make sure all contractors you hire are dependable and have a good reputation with suppliers and other contractors.
- For the best bids, take a comprehensive materials list to suppliers for quotes.

In This Chapter

- ◆ Planning the bathroom remodeling job
- ◆ Making sure you get the necessary permits
- ◆ Following building codes
- ◆ Getting the bathroom ready for remodeling

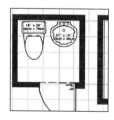

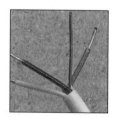

Chapter 9

Preparation and Permits

Ready to start tearing things out? Hold on! There are two additional important steps: preparing the bathroom for remodeling and making sure you have the needed building permits.

This chapter covers how and when you'll need a building permit for remodeling your bathroom as well as how to follow local codes. In addition, it begins the remodeling process by showing you how to prepare for remodeling your bathroom. You'll be getting ready for remodeling and taking care of any final problems.

The Remodeling Process

Of course, you're not the first homeowner to remodel a bathroom. That's good. That means there's a proven process for remodeling that you can follow to reduce time and costs. Whether you're working with a general contractor, hiring subcontractors, or doing all the work yourself, you can put the remodeling process or sequence to work for you.

Depending on how much of the work you do yourself, you may need to become familiar with one of the building codes, such as the International Residential Code (IRC), which combines regional building codes from across the United States into a single construction specification. You can buy copies of the IRC through your local building inspection department, larger bookstores, or online.

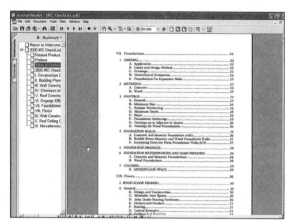

The International Residential Code (IRC) checklist is published by the International Code Council.

Remodeling Sequence

The sequence of remodeling depends on whether you're doing a bathroom facelift, makeover, or addition. Facelifts don't require moving any major components, but makeovers often do. Additions require new construction. Here's the typical remodeling sequence for larger jobs:

1. Structural changes (foundation, walls, subfloors, doors, windows)
2. Rough utility changes (wiring, plumbing, HVAC, cable)
3. Insulation
4. Ceiling and wall covering
5. Painting
6. Electrical fixtures and lighting
7. Cabinets, shelves, and countertops
8. Flooring
9. Plumbing fixtures
10. Décor

Chapters 10 through 19 will take you step-by-step through this process on a relatively direct path. If you're installing a new bathroom where none existed you'll begin with structural changes. If you are only doing a facelift you may begin with painting and move through

electrical fixtures, cabinets, flooring, plumbing fixtures, and décor.

Contractors and Suppliers

Unless you plan to do all the work yourself, remodeling a bathroom requires contractors, subcontractors, or hired labor. Chapter 8 covers hiring remodeling experts.

As a quick review, a contractor or general contractor is a construction manager and often an experienced tradesperson (carpentry, electrical, plumbing) as well. A subcontractor typically works for a general contractor by providing a specific trade or skill such as electrical or plumbing service. Alternately, you, the homeowner, can hire a subcontractor directly to tackle jobs you don't want to or feel qualified to do yourself.

Remember, if you have the needed skills but need some help you can hire labor. It may be a handyperson, a college student, or even a relative or friend to help you with the physical aspects of remodeling. In any case, part of preparing for remodeling your bathroom is making sure you have or can hire the skills and labor needed to do the job quickly, efficiently, and on budget.

For smaller remodeling jobs, spending a weekend morning buying materials is sufficient. However, adding a bathroom to your home typically requires more materials—and that means you should shop around for the best prices and delivery options. You may be fortunate and find a single supplier nearby that can furnish you with all lumber, utility, and remodeling materials needed at reasonable prices. In most cases, however, it will take two or three suppliers to get everything you need.

Construction Codes and Building Permits

Like it or not, larger remodeling projects require that you follow local building codes and pay for a building permit. Actually, it's for your own good. Codes are standards, and permits are a way of tracking remodeling done to a specific structure. An inspection is the physical review of a project by someone with knowledge of building codes and regulations.

Potty Training

Who actually issues building permits for your residence? It may be a township, city, county, state, or no one. To find out, get the legal description of your home's location (included on the paperwork you signed when you purchased the home) and take it to the local city or county. They can tell you who has jurisdiction. If they do, they also can tell you whether you need a building permit for your project.

Permits typically aren't required where there are no structural changes. Bathroom facelifts and some makeovers may not require a building permit. Check with your local authority.

Building Plans

It all starts with a building plan, a detailed drawing of exactly what you (or your contractors) will be doing to change an existing structure. It will include a drawing of the structure as it is today as well as the planned changes. Depending on how extensive the remodeling is, building plans will include some or all of the following:

- ◆ Framing plan
- ◆ Floor plans
- ◆ Wiring plan
- ◆ Plumbing plan
- ◆ Interior elevations
- ◆ Section drawings
- ◆ Specifications list
- ◆ Materials list

A trip to your local building department will tell you what permits will be needed, what plans you'll need to supply, and what building codes you must follow to remodel your bathroom.

Building Codes

As I mentioned in Chapter 4, building codes are the standards set up by someone somewhere to seemingly make it more difficult to build your house the way you want. Actually, the codes are designed for the health and safety of you, the residents of your house, neighbors, and anyone who buys the house in the future.

So does every county, city, town, and berg have its own building code? Kinda. Actually, most adopt a standard building code with maybe some modification depending on local needs and building practices. Building codes were developed as reactions to fires, earthquakes, hurricanes, tornadoes, and other disasters to minimize future damage to buildings. That means the codes required in earthquake-epicenter Los Angeles will certainly be different than those in rural Montana.

Until recently, building standards were regional. The western United States used the Uniform Building Code or UBC, the southern states typically followed the Standard Building Code or SBC, and the others used the National Building Code or NBC. Beginning in 2003, the codes are merged by the International Code Council into a single building code for the United States.

Building Permits

A building permit is an official document provided by a local building jurisdiction that permits you to build or remodel *if* you do so following approved plans and local building codes.

Do you need a building permit? Local building codes in various locations don't always agree on what requires a building permit and what doesn't. However, here are some typical home remodeling jobs that *don't* require a permit:

- Storage and tool sheds
- Fences under 6 ft. in height
- Painting, papering, and finish work
- Window awnings and shade structures
- Prefab swimming pools

Most jurisdictions require copies of your building plans and assurances that you will follow local building codes during the project. To enforce compliance, building departments will require that the project is inspected by an employee of the department, called a building inspector, and approved or signed-off at specific stages of the project. If you continue construction without the needed sign-off the inspector will probably require that you remove materials so the prior work can be inspected. Ouch! In the long run it's best to find out what permits are required, what codes need to be followed, and to allow inspections as required.

If you've hired a general or remodeling contractor to do most or all of the work, he or she will take care of the building permits for you and ensure that codes are followed. That's what contractors get paid to do.

What plans and paperwork are you going to need before you can apply for a building permit? Here's a typical list for new construction and larger remodeling jobs:

- **Plot plan** of the entire parcel with all existing and proposed structures
- **Floor plan** with the location, size, and use of each room, location and size of windows and doors, location of electrical outlets and subpanels, and location of plumbing and heating fixtures

- **Foundation plan** with all dimensions including exterior and interior footings, stem walls, pier blocks, and foundation support, and including footing depths, rebar, and anchor bolt locations
- **Elevation plan** of the finished exterior including all openings, siding material, original and finished grade, roof pitch, and roofing materials
- **Framing plan** for floors and roof including lumber grade, floor girder size and spacing, floor joists, wall studs, ceiling joist, and roof rafters and/or trusses
- **Cross-section plan** showing all primary structural elements from the foundation to the roof including heights and clearances
- **Signatures** of plan designers and engineers as required by code
- **Other stuff** including structural and engineering calculations, soil reports, and permits by other agencies as required by code and local building officials

First, you'll complete a Building Permit Application form and submit it with all the stuff to the appropriate building department. In most cases, you'll need a specified number of copies. Of course, there will be fees required for the time and effort to consider your building plans. Some fees may be due when the application is filed. Others may be due when the permit is issued. You may also need to pay any special levies, such as fire or school district, at that time.

Next, all the plans will be reviewed by the building department staff for compliance to requirements. Some departments will call and ask questions they may have. Others will just shoot the plans back to you and say "no" or "start over." If so, press them to find out exactly what needs to be changed to get the plan approved for construction.

The staff reviewer will make a recommendation to the department manager or whomever else has the power to say yes or no. Unless the remodeling project is controversial (such as lots of neighbors complaining), the manager will follow the staff's recommendations (so be *nice* to the staff!)

Following are typical inspection points for a single-family residence construction and larger remodeling projects. Bathroom additions might require all of these while makeovers only the final few:

♦ **Foundation inspection** after footing excavations are done, forms are built, and rebar is in place (depending on the type of foundation)

♦ **Slab or underfloor inspection** after the concrete slab, if any, or the subfloor is installed

♦ **Frame inspection** after all framing, bracing, and roof sheathing are installed and required utility rough-ins are done

♦ **Wall inspection** after exterior sheathing and drywall or lathing is installed, but before plaster or tape is installed

♦ **Final inspection** after the remodel is completed and ready for use

♦ **Other inspections** may be required by the building department depending on the type of construction and the plan's complexity; the building department will notify you in advance of additional inspections

In most cases, inspections must be "called for" prior to the inspector visiting the site. Find out how much advance notice is needed. Some will be able to inspect the same day while others may require a week depending on how busy the inspectors are (and how much clout the builder has!).

Potty Training

Your bathroom remodel might not require all of the inspection steps listed here. In fact, if you are simply giving the room a facelift you might not need *any* building permits or inspections. However, extensive remodeling, especially those that change the home's exterior as the neighbors see it, can require more details and more paperwork.

Preparing for Construction

Congratulations! You've been issued the appropriate building permits and know what codes you need to follow. You can now begin preparing for remodeling your bathroom.

What's the first step? Getting the old bathroom ready. That means removing any existing components as needed, exterminating any insects or other critters, and checking for radon, asbestos, carbon monoxide, and other health hazards.

Removing Existing Components

If your bathroom has walls that you won't be using, they will need to be removed before remodeling begins. Make sure the walls are not load-bearing, supporting the floor above. If you need to move a supporting post, consider hiring a contractor to do the work for you as there may be issues that can jeopardize your home's structural integrity. I'll give you more specifics in Chapters 10 through 12.

Insects

If you need to tear into walls, this is a good time to also take care of any insects or other pests that have or may soon damage your house. Contact a licensed pest exterminator to inspect the room and make recommendations for pest eradication. You also may need to make some repairs before continuing.

Radon

Radon is an odorless gas released when traces of uranium in the ground decay. Outside the home radon quickly dissipates in the atmosphere. Underneath a home radon gas can build up and become a health hazard, especially to those who smoke cigarettes or have respiratory problems.

Fortunately, you can buy a radon test or alarm kit at larger building materials outlets and use it to find out if hazardous amounts of radon gas (4 picoCuries per liter of air) are present. Instructions with the testers and alarms will tell you what to do next.

Heads Up!

Houses built before 1978 might contain lead paint, which can be hazardous if chips are ingested, especially by children. If you suspect that an area you are remodeling has lead paint, contact your local public health department for information on how to collect it and have a sample tested.

Asbestos

Asbestos, unfortunately, isn't as easy to detect. Asbestos was used in many building materials until about 25 years ago when the potential health hazards were identified. It was used in insulation, flooring, and other materials. Only airborne asbestos poses a health risk. That means removing some building materials in older homes can release asbestos particles into the atmosphere and potentially cause health problems.

Contractors trained and licensed for asbestos removal may be the best option if you suspect that your old bathroom has asbestos. Check your local telephone book under "Asbestos Abatement & Removal Service" or "Asbestos Consulting & Testing" to learn how to have your home inspected for this hazard. Depending on state laws, homeowners *might* be able to legally remove a small amount of asbestos material without a permit.

The Least You Need to Know

◆ Remodeling is a proven process with specific steps that are successfully completed by do-it-yourselfers every day.

◆ Make sure you visit the local building department soon to find out what permits are needed and which codes must be followed.

◆ Building plans are your map; comprehensive plans make the trip easier and more trouble-free.

◆ Before starting a bathroom remodel, make sure you remove any existing pests or hazardous materials.

In This Part

Part 2

Remodeling Your New Bathroom

Okay, you've put it off long enough. It's time to start making a mess!

The first part of this book showed you how to analyze, plan, design, pay for, and prepare for your new bathroom. This part takes you to the next level by showing you how half baths, full baths, and extended baths are remodeled.

Don't be nervous. Sure, you still can call in a contractor or a specialist to do some or all of the work. Or you can develop some "sweat equity" by tackling some or even all of the tasks that need to be done. In fact, if you've done a little remodeling before this part may have all the directions you need to get the job(s) done.

So circle a date on the calendar—preferably one in the next decade—and put your plans into action. Start remodeling your bathroom!

In This Chapter

- ◆ Comparing your half bath design to common designs
- ◆ Considering upgraded and unique half baths
- ◆ Planning the installation of half bath components
- ◆ Basics of installing half bath toilets and lavatories

  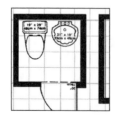  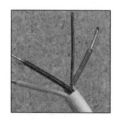 

10

Remodeling Half Baths

One of the easiest and least expensive bathroom remodeling projects is modernizing a half bath. A half bath facelift, for example, can be completed in a weekend for under $1,000. A makeover can double or triple the cost with significant results. Adding on a half bath can set you back many times more money, but give you valuable results.

This chapter shows you how to remodel smaller bathrooms, called half baths, including actual projects and introducing some of the specific steps. Additional instructions are offered in Chapters 13 through 19 as each bathroom system or component is explained and illustrated.

Popular Half Bath Designs

As you learned in Chapter 1, a half bath is called so because it typically only contains a toilet and sink or lavatory. It doesn't have a shower, tub, or other major plumbing fixture. In addition, a half bath usually is in a smaller room, commonly one of less than 25 sq. ft., though a half bath can be part of a larger room such as a utility or storage room. In fact, some folks extend an existing laundry room with a half bath.

Because half baths are relatively small it is perceived that there is little you can do to them. That's not true. Chapters 1, 2, and 3 introduced some ideas for half bath facelifts, makeovers, and additions. In Chapter 5 you designed your bathroom remodel. Here are some actual half bath remodeling projects based on basic, deluxe, and unique designs.

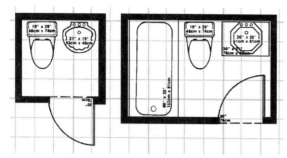

The primary difference between a half bath (left) and a full bath (right) is the addition of a bathtub or shower.

Basic Half Bath

Because the typical half bath is less than 25 sq. ft. in size there may only be one or two variations to the layout of the two primary plumbing fixtures: toilet and lavatory. Layout is even more limited if access includes two doors or the room is located under a stairway where headroom is at a premium. Once you have your basic half bath design, here are some project guidelines:

Half baths typically are small and include just a toilet and lavatory.

◆ Make sure that water, electrical, and other utilities are shut off to the room before working.

◆ Remove old components that will be replaced.

◆ Prepare all surfaces, repairing as needed.

◆ Paint or resurface all walls and ceilings.

◆ Install new moisture-resistant flooring as needed.

◆ Install a two-piece, floor-mounted, tank-type toilet of vitreous china.

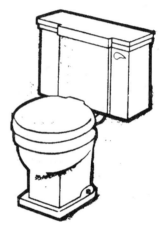

Two-piece, floor-mounted, tank-type toilet.

Potty Training

Some building codes require that any new or replacement toilet must be a 1.5 gallon water-saver toilet. Older models had 3 and 5 gallon tanks. Make sure you know what local codes require before selecting a new toilet. If in doubt, ask your materials supplier.

◆ Install a vanity base cabinet and vanity top.

◆ Install an 18 in. round or oval porcelain-coated cast iron or similar sink, also known as a lavatory.

◆ Install faucets and other lavatory fixtures. (Remodeling contractors remind you that you get what you pay for. Don't go cheap on faucets and fixtures. Today's special price may be tomorrow's repair!)

◆ Install a wall-hung medicine cabinet.

Waste Lines

You can save money on your medicine cabinet by purchasing an inset model that fits between wall studs. It's cheaper; the sides don't have to be finished because they don't show.

◆ Install towel bars, toilet paper dispenser, storage shelves, and other accessories. (Wall anchors just aren't strong enough to hold up a towel bar, so install toggle bolt anchors instead.)

Chapters 13 through 19 offer the specific steps for each of these tasks. This action list is for initial planning.

Deluxe Half Bath

A deluxe half bath is one that includes various upgrades in plumbing fixtures and decoration to make it more functional, attractive, and efficient. Whether your deluxe half bath project is a facelift, makeover, or addition, here are some project guidelines:

◆ Make sure that water, electrical, and other utilities are shut off to the room before starting.

◆ Remove old components that will be replaced.

◆ Prepare all surfaces, repairing as needed.

Heads Up!

One of the most common repairs remodeled bathrooms need, especially smaller ones without adequate ventilation, is subfloor repair. Moisture collects in the room and softens wood components, especially the flooring around the base of the toilet. Moisture also damages walls near sinks. Make sure you check and repair these locations as needed.

◆ Paint or resurface all walls and ceilings.

◆ Install new moisture-resistant flooring such as ceramic tile or quality vinyl composition tile (VCT).

◆ Install a one-piece, floor-mounted, tank-type toilet of vitreous china.

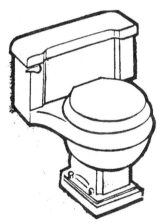

One-piece floor-mounted toilet.

◆ Install a deluxe vanity base cabinet and single-piece vanity top or individual ceramic tiles.

◆ Install a deluxe 18 in. or larger round or oval porcelain-coated cast iron or similar sink, also known as a lavatory.

◆ Install deluxe faucets and other lavatory fixtures.

- ◆ Install a lighted plate glass mirror above the vanity.
- ◆ Install deluxe towel bars, toilet paper dispenser, storage shelves, and other accessories.

As covered in Chapter 6, a deluxe version of a half bath can add 50 percent or more to the final cost of the remodel. If your budget is a little smaller, adding your labor or your bargaining skills to the project can give you a deluxe bathroom on a basic budget.

Unique Half Bath

Even small half baths have options. There are things you can do as you remodel your bathroom to make it seem larger as well as get the most from each dollar and hour you spend on the project:

- ◆ If you are simultaneously remodeling other bathrooms in your home, consider moving fixtures from one bathroom to another.
- ◆ Remember that you can resurface wood and even porcelain with unique colors and textures.
- ◆ Check local salvage yards for materials that may make your half bathroom more unique.
- ◆ Consider bright colors to liven up an otherwise dark half bath stuck under a stairway or other sheltered location.
- ◆ To make a small bathroom look larger add mirrors to walls.
- ◆ Once done with a basic or deluxe half bath, add trim and accessories to make it unique (nautical motif, wall mural, hanging plants, and so on).

Half baths are relatively easy to remodel, typically within the budget and the skills of most do-it-yourselfers. Now that you've seen some common designs and the steps to remodeling them, let's take a closer look at the biggest part of bathroom remodels: plumbing.

Plumbing a Half Bath

How much you'll need to know about plumbing depends on what type of remodel you're doing. A facelift may not require much more than replacing a faucet or two while an addition will require that you install pipes in walls.

Chapter 13 will cover plumbing in more detail, but as you plan your half bath remodel job you need to know some of the fundamentals of plumbing. Here's a summary of plumbing basics, selecting fixtures and cabinets, and other elements as they relate to the half bath.

Plumbing Basics

Plumbing or the plumbing system is made up of components that distribute the water supply. In residences it includes the pipes, valves, fixtures, traps, vents, drains, and other components. It gets water from the city water utility or a well or spring on the property and delivers the waste to a city sewer utility or a septic system on the property.

The plumbing system for a half bath is relatively simple because the fixtures typically are a toilet and a lavatory or sink for washing. Both of these fixtures require water and waste lines. In addition, the lavatory needs both a hot water and a cold water line. Toilets only require cold water lines. A bidet typically has both a hot and cold water source.

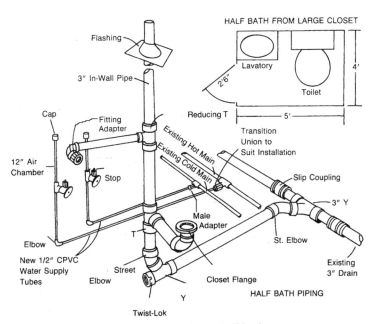

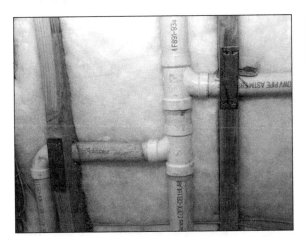

Typical piping for a half bath.

DWV (drain-waste-vent) lines are installed in walls to allow egress for waste water from bathroom fixtures.

New construction uses a combination of PVC (polyvinyl chloride), PEX (cross-linked polyethylene), and copper materials for ease of installation.

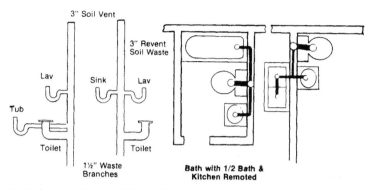

Typical piping for a half bath that's near but not adjacent to a kitchen or other established plumbing system.

Half baths often have both primary fixtures on a common wall; that is, they are side by side. This arrangement is more efficient (and less costly to build) because the supply and drain pipes are in a single wall. The lavatory's drain pipe takes water away from the basin and dumps it in to the main vent stack. The toilet dumps waste into the main vent stack as well.

Selecting Fixtures

As you discover when shopping for plumbing fixtures, there is a wide variety to choose from. Your remodeling plans will include sizes and dimensions, but may not be specific otherwise. Here are some guidelines for selecting half bath plumbing fixtures:

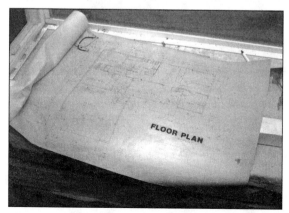

Your remodeling plans will include sizes and dimensions for installing all fixtures.

◆ Vitreous china is still the most common toilet material but various plastics are increasingly popular.

◆ Toilet tanks can be included in the toilet (one piece), attached to the toilet (two piece), or separate with the tank attached to the wall (Victorian).

Wall-mounted toilet.

◆ Tank widths vary from 20 to 24 in.

◆ Toilets project 26 to 30 in. from the wall.

◆ The nearest fixture to the toilet side should be no less than 15 in. from the center of the toilet.

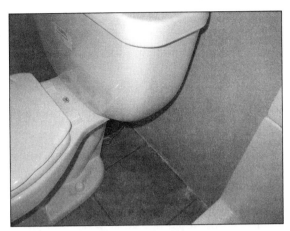

Make sure that the toilet is at least 15 in. from adjoining fixtures.

Installing a toilet too close to adjoining fixtures can make use difficult.

◆ There should be a clear space of at least 21 in. deep in front of the toilet.

◆ Low-flush toilets require less than 2 gallons per flush, about one third that of older toilets.

◆ Pump and pressurized toilets are available for basements and other below-grade applications.

◆ Sink sizes vary from 12 to 30 in. or larger.

◆ Sink shapes can be round, oval, square, or rectangular.

◆ Freestanding sinks are increasingly popular, designed to sit atop a cabinet.

◆ Pedestal sinks take up less space than cabinet mounted sinks, but don't offer storage.

Potty Training

Freestanding and pedestal sinks also are available in wall-hung versions that give them more structural support.

◆ Purchase your toilet, sink, and cabinets before starting the remodel to make sure that the model you want is available.

As mentioned, you'll learn much more about plumbing and fixtures in Chapter 13. The information and tips here are for half baths.

Selecting Cabinets

Small half baths require smaller cabinets. A minimum width for a standard lavatory or sink cabinet is 2 ft. 6 in. or 30 in., though narrower versions are available. A freestanding sink is popular if there is sufficient storage space elsewhere in the room or nearby. Otherwise, choose a standard cabinet that gives you a width of 30 in. for the toilet area. Here are the specifications for smaller bathroom cabinets:

◆ Sink base or vanity cabinets typically are 29 in. high (excluding the counter top).

◆ Modular base cabinets are available in 6-in. width increments up to 60 in.

- ◆ Double bases (two doors or sets of drawers) are 30, 33, 36, or 42 in. wide.
- ◆ Single bases are 12, 15, 18, or 21 in. wide.
- ◆ Standard bathroom vanity cabinetry depth is 21 in.

Depending on the layout of your half bath you may have chosen to install an overhead cabinet for storage. If so, make sure that it offers sufficient head clearance so no one gets bonked. Alternately, install a cabinet above the toilet for storage to take advantage of otherwise wasted space.

If you're using existing cabinets, remember that you can resurface them either with veneer or with household paint before reinstallation. A popular facelift is to resurface the cabinet and replace the counter top and sink.

> **Waste Lines** _____
> You can have the European look and save money on your bathroom remodel by buying unfinished cabinets and finishing them yourself. Paint the cabinet face frame white or a light color, then use a clear finish on the natural wood door and drawer faces. Cheap elegance!

Selecting Accessories

Bathroom accessories shouldn't be an afterthought. They should be part of your bathroom remodeling plan. Make sure you know where the toilet roll holder, towel rack(s), mirrors, lighting, and other components will be installed before you begin. In fact, it's a good idea to buy them even before you start the remodel so you are sure you'll have them when ready. You also may refer to them for sizing and placement questions as you remodel.

> **Waste Lines** _____
> Running out of space around your bathroom toilet? Install a recessed toilet roll holder for slightly less protrusion into the room.

If you are installing accessories such as grab bars for the toilet, get a copy of the ADA Guidelines for locations (see Appendix B). Because some components require reinforcement you should know exactly where they are going well in advance so you can make any needed modifications to the walls. Here are some guidelines:

- ◆ Grab bars should be 1¼ to 1½ in. thick and mounted so there is at least 1½ in. space behind the bar for gripping.
- ◆ Side bar near the toilet should be at least 12 in. from the back wall and 42 in. long.
- ◆ Rear bar behind the toilet should be at least 24 in. long (36 in. is preferred) and extend toward the access side of the toilet.
- ◆ Grab bars should be mounted horizontally 33 to 36 in. above the floor.

Installing Toilets and Lavatories

Toilets and lavatories (sinks) are relatively easy to install, even for the first-time do-it-yourselfer. If your half bath remodel only requires that you replace a toilet and/or a lavatory, you may not need to read beyond this chapter. If your remodel requires adding or moving pipes or getting farther into the plumbing system, you will need the more extensive instructions in Chapter 13.

To install a toilet or lavatory, you'll need the chosen plumbing components, various readily available tools, space to work, and some time. How much time? Depending on what preparation is required (removing the old tank, subfloor repair) the job could take a full day. However, if you are simply installing a new toilet where the old one went, plan on doing the job in 2-3 hours. If you're prepared and have a little plumbing experience, you can get it done in less than an hour, like the plumbers do.

Tools and Materials

Materials you need include the toilet itself (one- or two-piece), toilet wax ring, and assembly hardware. You'll also need a toilet flange and water lines available from your hardware or plumbing store. If you're installing a new lavatory (cabinet and sink) or freestanding sink, you'll need the components standing by.

Toilet flanges are installed 12 in. from the wall.

Tools are as simple. You'll need a wrench (adjustable, ratchet, or basin), a screwdriver, and any tools required for installing or replacing the water supply line (Chapter 13). To install the lavatory you'll only need adjustable wrenches and a screwdriver unless you also are installing or moving supply lines (Chapter 13).

Remember the adage: If all else fails, read the directions. That is, make sure you pick up and read the manufacturer's direction sheet for fixtures you buy. Some are unclear or misleading as they try to cover too many models or aren't up to date. However, most instructions are helpful. Reading them over in advance may save you time and frustration.

Preparation

Preparation depends on what you're doing. If you're installing a new toilet where none has gone before, Capt. Kirk, there is little prep work. However, if you are replacing an old toilet, moving it and repairing the subflooring, the prep work may take longer than the installation.

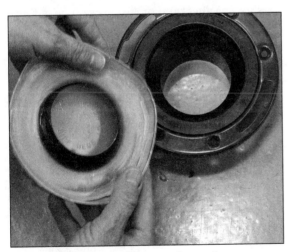

Typical toilet flange and wax ring.

Potty Training

Don't forget to measure! Take a measurement from the toilet flange floor bolt to the wall behind the toilet. Most toilet flanges are installed 12 in. from the wall.

Preparation often requires moving supply and drain lines within walls.

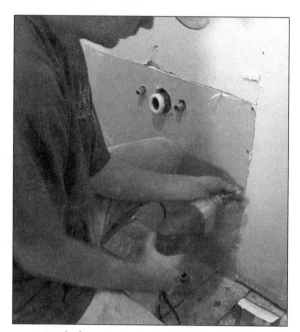

Instead of tearing out a large section of wall to access pipes, you can cut a smaller hole and, when done, replace the removed drywall.

The wall and plumbing are then ready for installing the lavatory and fixtures.

In this chapter, we're only installing a new toilet and a lavatory. Refer to subsequent chapters for specifics on more extensive preparation.

Make sure the toilet flange is in the correct location before installing flooring and the toilet.

Toilet Installation

What's the difference between a single-piece and two-piece toilet other than the obvious? Cost! A single-piece toilet is more difficult to manufacture so it will cost more to buy. In addition, it weighs more than the individual components of a two-piece toilet so it's more difficult to move. However, installation is approximately the same for either model.

Here are the steps to installing a replacement one- or two-piece toilet:

1. Make sure that water source is turned off and that the area is ready for installation of the new toilet unit.

2. Turn the bowl upside down and install a new wax ring and sleeve on the toilet horn (directions are on the ring box).

3. Carefully position the toilet over the toilet flange and lower the toilet into position. This is a much easier job if you have a helper.

4. Press down on the toilet bowl to compress the wax ring, then install and firmly tighten the washers and nuts.

5. If installing a two-piece toilet, install the tank next. Install the spud washer (included) and center it over the inlet opening, then install and firmly tighten washers and nuts.

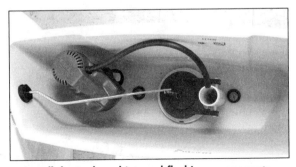

Install the tank and internal flushing components.

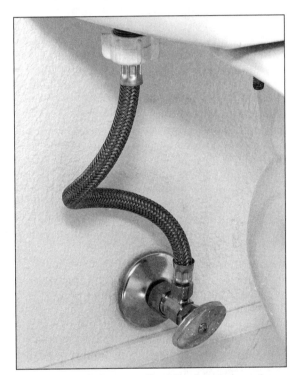

Potty Training ⎯⎯⎯⎯⎯

If you don't have much room in your small bathroom, assemble a two-piece toilet outside of the room and carefully bring it in for installation.

6. Attach the water supply line to the tank. If necessary, refer to Chapter 18 on installing fixtures to change the water supply line.

Water supply lines are flexible tubes that can be installed with a small wrench.

7. Mount the toilet seat.

8. Turn on the water supply line, let the tank fill, and flush to test.

A fully installed two-piece toilet.

Lavatory Installation

A lavatory is a built-in wash basin, typically installed in a cabinet. A freestanding wash basin also is called a lavatory—as is the room in which they are housed. In this book, a lavatory is where you wash your hands. Here's how to install the two primary lavatories.

Installing a built-in lavatory:

1. Attach the vanity cabinet to the wall, making sure that the sink will be above the drain line protruding from the wall.

2. Place a level on top of the cabinet and insert thin wooden shims under the vanity as needed to make sure that it is level.

Potty Training

Most bathroom vanities don't have backs, only a supporting cross-member. If your cabinet has a back, determine the correct location on the wall for the vanity, then mark the location of the pipes and cut holes in the back of the vanity so they can pass through the back once installed.

3. Find the studs in the wall (using a stud finder or tapping on the wall for non-hollow areas) and attach the cabinet to the wall with screws.

4. Install the vanity top following directions that came with the unit. If building your own top, install a ¾-in. thick cement board base over the top then install the tile, Formica, or other final surface following manufacturer's directions or Chapter 17.

Install the vanity top over the cabinet base.

The sink opening in the vanity top should be directly above the water supply and drain lines exiting the wall.

5. Install the sink, fastening it to the cabinet from the underside using a screwdriver or from on top with silicone caulking depending on the manufacturer's recommendations.

6. Follow manufacturer's directions or Chapter 18 for installing the new faucet(s) and attaching the water lines.

Fully installed sink, water supply lines, and drain trap system.

Installing a freestanding lavatory:

1. Set the base and pedestal in position to check for fit. Most freestanding lavatories are two-piece with a basin and a pedestal.

2. Depending on the model, mark the floor where the pedestal will be attached to the floor. Also mark the wall as needed for installation of the basin. Follow manufacturer's directions as there are many types and installation methods.

3. Remove the basin and pedestal and set the aside. As needed, mark and drill holes in the floor for installation of the base.

4. Install any mounting hardware needed for attaching the basin to the wall. If necessary, attach the base to the pedestal. (Some models are connected after installation.)

5. Install the base and pedestal in position.

6. Connect water supply and drain lines as needed. Follow manufacturer's directions or Chapter 18 for installing the new faucet(s) and attaching the water lines.

The Least You Need to Know

◆ Half bath remodels are relatively easy as most only include a toilet and a lavatory.

◆ Make sure you get all needed remodeling components before you start the job to make sure you have what you need.

◆ For practicality and to meet building code requirements, be sure to install all bathroom components with the proper clearances.

◆ Bathroom toilets and lavatories are relatively easy to install once the preparation work is complete.

In This Chapter

- ◆ Planning a full bath remodel
- ◆ Selecting bathtubs and showers
- ◆ Choosing cabinets and accessories
- ◆ Installing bathtubs and showers

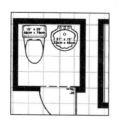

Remodeling Full Baths

The majority of modern bathroom remodels is of full baths, ones with a tub and/or shower. With so few major components you'd think that there would be few variations. However, remodeling suppliers have come up with a wide variety of designs, models, sizes, and features for every bathroom component including the single-purpose bathtub.

If you are remodeling a full bathroom, this chapter will give you an overview of your project and help you decide how best to tackle the job. If you're doing a facelift or a makeover and you're already handy, you may only need to refer to this chapter. If you want to learn more about plumbing, electrical, and other bathroom systems you can next refer to Chapters 13 through 19 that go into greater detail.

Popular Full Bath Designs

As you learned in Chapter 1, a full bath includes a toilet, lavatory, and a bathtub, and is typically 50 to 100 sq. ft. in size. Bathrooms with a shower *instead* of a bathtub are sometimes called three-quarter baths, but they really are full baths.

That's the basic design. There are numerous variations including a full bath with two toilets, a toilet and a bidet, two separate lavatories, a combined bathtub and shower, a separate bathtub and shower, and many other versions. Once you go beyond the toilet/sink/tub combination the bathroom typically is considered a master bath or extended bath (see Chapter 12).

A full bath has a toilet, lavatory, and bathtub (or shower).

Basic Full Bath

Because the typical full bath is twice the size of the typical half bath there are many more variations for what you can install where. Depending on your existing or planned plumbing system, the toilet, for example, can probably be installed in one of a number of locations. You have more options.

Components of a basic full bath.

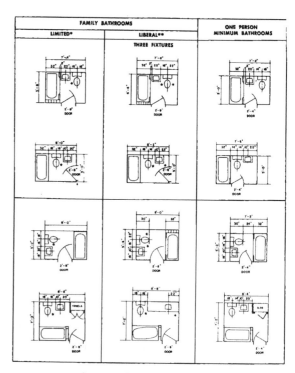

Twelve popular full-bath designs.

Of course, you also have limitations. For economy, maybe you planned to have all water fixtures on the same wall or on adjacent walls so the plumbing run is short. Here are some tips on planning the remodel of a basic full bath:

◆ If your budget allows, spend a hundred dollars or more on upgrading the sink, tub, or faucets; shopping around can often get you wine on a beer budget.

◆ Remember to verify that the subfloor is in good condition before installing new flooring and fixtures.

◆ If you're uncomfortable with installing new plumbing, hire a plumber to instruct or assist you on the best ways to install.

◆ When buying larger components such as a tub/shower combination, ask the dealer for a copy of the installation instructions that you can study before purchasing.

- If you are installing a new tub or shower unit, carefully measure clearances of doorways in your home in the path to the remodeled bathroom.

- Make sure you have all major components on hand before preparing the room for the remodel.

Deluxe Full Bath

If you're going to spend all that time and money on a remodel it makes sense to consider dressing it up into a deluxe full bath. As discussed in Chapter 6, the typical budget for a basic full bath remodel is about $6,000 and the deluxe version comes in around $9,000. These numbers, of course, depend on how extensive the remodel is and whether you're adding on a room or just giving an existing one a major facelift.

You can extend the value of each remodeling dollar with a few tips from the pros:

- Reuse what you can; instead of replacing perfectly good plumbing, reuse it and apply the money saved to upgrading other components.

- Make sure you install ½-in. thick water-resistant sheetrock on any wall that may come in contact with water or condensation.

- Buy an extra 10 percent of any material, such as flooring or wall tile, that may require future repair or replacement.

- If installing your own floor tile, consider 12 in. square sheets of 1 in. square tiles as they are easier to trim and don't require a tile cutter.

- Plan your remodel so that all painting can be done at once, saving time and labor.

- To extend the perceived size of a bathroom—as well as its functionality—install a full-length mirror on the back of a door.

- Before starting the remodel, use a level to make sure that all wall and floor surfaces are as level, plumb, and perpendicular as possible.

Heads Up!

If you're installing your own floor tile over an existing subfloor, make sure the subfloor is strong enough to stand up to future use. Consider adding or replacing subflooring and installing waterproof treatments so you don't have to do the job again.

Unique Full Bath

If your full bath is "unique," it may have recycled fixtures, an unusual motif, or other components that give the room character as well as function. Unique full bathrooms utilize colors, textures, and unusual components such as *gravity-flush toilets* or *incineration toilets*. You've planned these unique elements into your new full bath. Here's how to implement them:

Bathroom Words

Most toilets use gravity and water pressure to clear the bowl. New **gravity-flush toilet** designs use taller, narrower tanks to increase the flow in low-flow models. **Incinerator toilets** use fire to burn waste safely.

- Make sure that you have all components, especially unique ones, on hand before you begin preparing the room for remodeling.

- Know that functioning components that are unique (tub, toilet, lavatory) may require a special review by the building department before it is allowed to be installed in your new bathroom.

◆ If you or someone is painting a mural in your unique full bathroom, consider having it done once the old components are out and before the new ones are installed.

◆ Remember that tiles require a filler between them, called grout, and the grout subsequently will need a sealer to keep moisture from penetrating.

◆ Consider taking photos of your unique bathroom remodeling project for your scrapbook.

Plumbing a Full Bath

Plumbing a full bath is a little more complex than plumbing a half bath (see Chapter 10) because you are adding a major component: a bathtub or shower. However, think of the tub or shower as a big sink and you'll have the concept. The tub or shower has hot and cold water delivered and waste water drained away. Faucets control the flow and mixture of the water.

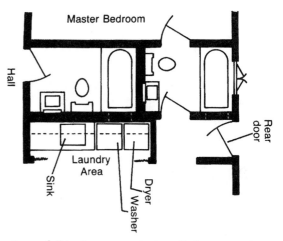

A new full bathroom can be installed near another bathroom or a laundry room to share plumbing.

The most difficult part of installing a new bathtub or shower may be getting it in place. Because they are bulky, your choices will be limited to units that you can get into the room. Fortunately, the market for remodeling is big enough that many manufacturers offer models that can come in the door. Make sure you know the unit's measurements (including shipping carton) before buying it.

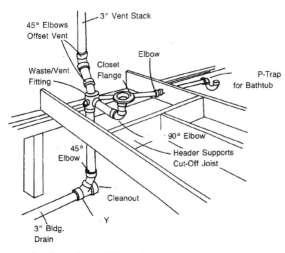

Plumbing for a full bath with toilet and bathtub.

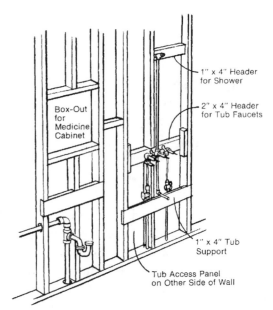

Plumbing components inside the wall of a typical full bath.

Selecting Bathtubs

There is a wide variety of bathtubs available of various designs, sizes, and materials. Which should you choose for your full bath remodel?

In some cases, the best tub is the one that is already in place. You may be able to repair or refinish it, saving time and money. If not—or if you're just tired of the old tub—all you have to do is visit a large building material or plumbing supplier to be amazed at what is available.

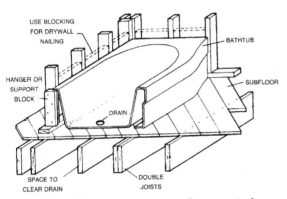

Structural components surrounding a typical bathtub.

Most modern bathtubs are made of fiberglass, solid acrylic, or coated cast iron. Fiberglass tubs are the least expensive, but the finish doesn't last as long unless treated with an acrylic topcoat. Solid acrylic tubs are lightweight and more durable—and more expensive. Most whirlpool and specially shaped tubs are made of acrylic. Cast iron tubs have been around the longest and will last the longest. They are coated with a porcelain enamel. They also are the heaviest and most difficult to install.

Bathtubs that abut a wall typically are 32 in. wide by 60 in. long. Elongated tubs are 72 in. in length and special tubs can be found in widths of 24 to 42 inches. Standard heights range from 12 to 20 in. with 14 in. the most common height.

Bathtubs can be installed on the floor, on a platform above the floor, or into the floor. However, building codes will require grab bars and steps or treads for anything other than standard on-the-floor installation.

Selecting Showers

Think of a shower as a stand-up bathtub. Many people prefer the shower over the tub for cleaning and relaxing, so many homes have both. The most popular shower is actually the tub/shower combination. It's especially popular in new homes and is installed even before other interior components because it is so big that the house is built around it.

Potty Training

A shower drain is 2 in. in diameter and a bathtub drain is 1½ inches in diameter. Showers need a larger drain for faster drainage because they don't have depth to hold excess water.

Prefabricated showers and tub showers are available as single units or as *components*, more popular with remodelers. Materials include fiberglass or acrylic. Showers also are built at site using ceramic tiles.

Bathroom Words

Component showers that are comprised of a shower base and separate walls are designed primarily for remodelers who must get the unit into an existing space. These units are commonly called *knockdowns* or *KDs*.

Prefabricated showers available at larger home improvement centers are ready to install.

Standard showers are 32, 36, or 48 in. wide with 36 in. being most common. Standard height is 73 inches.

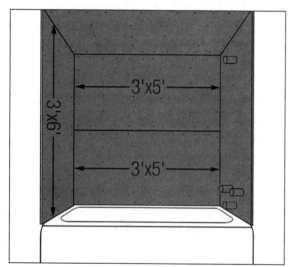

Measurements for a typical tub-surround shower.

Selecting Cabinets and Accessories

Full bathrooms typically have larger cabinets than half baths. In fact, if wall space is available your full bath can have additional full-height storage cabinets that serve as a minicloset.

As mentioned in Chapter 10, most bathroom floor cabinets are 29 in. high (excluding the top), 21 in. deep, and 30 to 42 in. wide. Extended bathrooms (Chapter 12) often include wider cabinets for two sinks. Make sure that the cabinets you select will fit through the bathroom door *and* be able to be turned into position. An alternative is to buy narrower cabinet components, such as two 30 in. cabinets, designed to be fastened together. Chapter 17 offers more information about installing cabinets in bathrooms.

Accessories for a full bath are similar to those for a half bath; there are just more of them. In fact, a full bath may have sufficient wall space for drying bars so that multiple towels can be hung at once. There are coordinated groups of accessories available at bathroom shops that can add luxury to your remodel project.

Installing Full Bath Fixtures

The basics of installing a toilet and lavatory were included in Chapter 10. What the full bath adds over the half bath is a tub, shower, or tub/shower combination. More extensive installation instructions including how to move pipes and tackle difficult plumbing jobs are covered in Chapters 13 and 18. Here, let's take a look at how to install a typical tub and shower.

You've designed your full bathroom and selected components. What else do you need to start the job? Tools, materials, and room preparation, covered below. How long will the job take? Prep time depends on the condition of the existing room, the complexity of the remodel, and your experience. Preparing the room for a new tub or shower could be just a few hours or a few days, depending on whether you need to move walls or plumbing. Obviously, installing a sunken tub will require more time.

The actual installation of the tub or shower including faucets should take less than one day. If you are new to remodeling, plan to do it over a two-day weekend to be safe.

Tools and Materials

Materials you need include the bathtub, shower, or tub/shower combo, plus the new faucets and drain, caulk, plumber's putty, dry-set mortar, and some wood shims for leveling the tub or shower.

> **Waste Lines** _____
>
> Make sure you get clear, readable installation instructions with the tub or shower you buy as some have special requirements. In addition, some manufacturers include single-use tools for alignment or installation that will save you buying an expensive tool.

Tools include plumbing wrenches, carpenter's level, tape measure, screwdrivers, utility knife, and a caulk gun. You also may need a saw for cutting plastic pipe or a copper tube cutter.

Preparation

Preparation is getting the location ready for the fixture. If you're simply replacing one built-in acrylic tub/shower with a newer version the prep may be as simple as removing the old unit. The process to removing a bathtub is:

1. Separate the tub from the walls with a utility knife and pry bar.
2. Remove the drain and any other fixtures within the tub.
3. Lift front edge of the tub from the floor with a pry bar.
4. Pull the tub away from the wall.

5. Remove the tub from the room. If necessary, cut a fiberglass or polymer tub up using a reciprocating saw for easier removal.

Removing a fabricated shower is relatively easy:

1. Remove all fixtures, including faucets and drains.
2. Remove the sidewalls from the walls with a utility knife and pry bar. If tile, you may or may not decide to remove all of the tiles depending on what will replace it.
3. Lift the front edge of the shower base from the floor with a pry bar.
4. Remove the shower base.
5. Remove the shower components from the room.

From this description you can figure out how to remove a tub/shower combination. Remove the shower walls first, and then the tub that also serves as the shower base.

Installing Bathtubs

Installing a replacement bathtub is relatively easy and, once prep work is done, should only take a few hours. However, a replacement tub is one that doesn't require any plumbing changes. You're replacing an old unit with a new one of the exact same size and fixture location. If you are installing a different size or style of tub the preparation becomes more complex—moving supply pipes and drains, moving walls, or replacing a shower with a tub. That's when you need the advanced plumbing tips offered in Chapter 13.

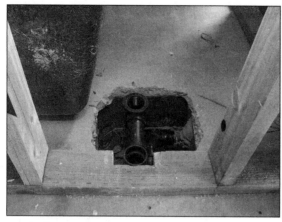

Bathtubs are slid into place over existing drain pipes.

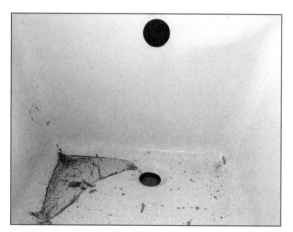

Make sure that the bathtub drain opening fits exactly over the drain flange.

Potty Training

Make sure you install the drain flange and any stopper mechanism before you completely enclose the bathtub.

Here are the steps for installing a replacement bathtub once the preparation work or other remodeling work is done:

1. Attach the tub rim to the wall studs, adding shims as needed for an even edge.

2. Attach the drain and overflow to the appropriate pipes using plumber's putty. Most drains screw into place and overflow covers are attached with two screws.

3. If the tub has shower walls above it, install them (below) before installing fixtures. If the walls above are to be tiled, install the tile (see Chapter 16).

4. Caulk or seal the joints between the tub and walls as needed to minimize water damage to surrounding surfaces.

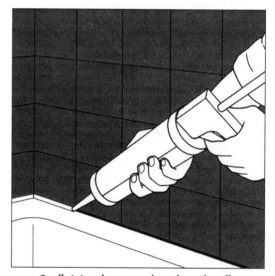

Caulk joints between the tub and walls.

5. Install the faucet(s) and spout (see Chapter 18).

Some bathtubs don't have mounting holes for fixtures, but instead have them above the tub rim and on the wall.

That's how the job goes. Depending on the complexity of your remodel, these instructions may be sufficient to replace your bathtub with a new one. If you are installing a whirlpool or other special tub, basic instructions are in Chapter 13. If the replacement is more complex, refer to the chapters in Part 3 on Advanced Remodeling.

Installing Showers

Like tubs, installing a replacement shower with a similar unit is a fairly easy job. Even replacing a fiberglass unit with a fiberglass-base/tile-wall shower is not much more difficult depending on whether you're moving the drain or the faucets. Instead of installing fiberglass walls you simply install tile ones (see Chapter 16).

Here is the process for installing a replacement shower panel kit:

1. Install the drain in the shower base (sometimes called the "pan") and tighten.

2. Apply dry-set mortar to the subfloor where the base will be installed.

3. Lower the shower base into place over the drain and place it in the mortar, adjusting the base as needed to make sure it is level. Allow the mortar to set for 8 hours or as directed by the manufacturer before continuing.

4. If included with the kit, install the shower dome following the manufacturer's instructions. Most domes require additional framing and lighting.

5. Install the shower door mounting strips to the frame for later installation of the door.

6. Accurately measure and cut holes in wall panels to accommodate faucets and other accessories.

Installed shower base.

Shower walls are installed after the base.

Measure and cut holes in wall panels to match up with existing pipes.

7. Assemble the shower panels. Many manufacturers use a corner clip system on the back corners.

Shower panels are installed directly to studs or to waterproof drywall called greenboard.

8. Apply waterproof panel adhesive on the bare walls where the shower walls will be installed.

9. Carefully slide the shower panel assembly into the alcove and press the panels against the walls.

10. Use padded scrap lumber to press the panels against the walls so the adhesive bonds tightly.

11. Install the faucets, shower head, and other finish plumbing components.

12. Seal the seams around the base with tub caulk.

Some shower heads are installed above the walls.

Match up existing faucets with the shower wall for a neat installation.

Seal the shower seams with caulk.

13. Install the shower door following the manufacturer's directions.

Again, for a tub/shower combo, install the tub then follow the instructions above for installing the shower walls.

Single-unit tub/shower combos are more difficult to install because they are large and may not fit through doorways.

The Least You Need to Know

◆ Depending on the complexity of your full bath remodel, the basic instructions in this chapter may be sufficient to get the job done.

◆ Make sure you select replacement components that not only fit, but also make your remodeling job easier.

◆ Cabinets come in a variety of sizes and designs that are relatively easy to install.

◆ Many homeowners have the skills to install a replacement bathtub or shower in a day or a weekend.

In This Chapter

◆ Developing your new extended bathroom project

◆ Choosing components for your extended bathroom

◆ Selecting tools and materials for the job

◆ Installing jetted baths, special showers, and saunas

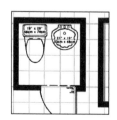

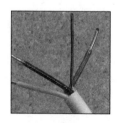

Chapter 12

Remodeling Extended Baths

Many people think of someday remodeling their old bath into a dream bathroom. Then they get the cost estimate and settle for a remodeled full bathroom. However, as you've learned in this book, doing some of the work yourself and shopping smart you can extend your remodeling budget and get more than you may have first expected.

This chapter shows you how to turn your plans for an extended bathroom—one with more space and more features—into a reality. You'll learn how to select specific major components as well as how to install them. Your dream can become a reality!

Popular Extended Baths

As you learned in Chapter 1, an extended bathroom is one of at least 80 to 100 sq. ft. that includes fixtures beyond the basics. Besides a toilet, lavatory, and tub or shower, it may have a spa, hot tub, sauna, steam room, special shower, larger fixtures, a dressing or powder room, a garden area or fountain, or other components that make the room more appealing for relaxing and longer visits. A master bath typically adds a shower and a second lavatory sink.

If you have enough room in your existing bathroom you can add many of these features and fixtures as a makeover. If sufficient room isn't available you need to make an addition. The specifics of how to do advanced remodeling is included in Part 3. This chapter will focus on easier tasks for the do-it-yourselfer as well as an overview for those who have chosen a remodeling contractor but still want to know what's going on.

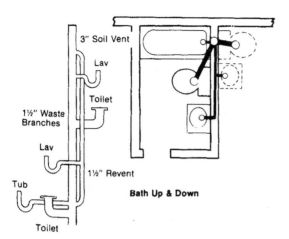

Consider installing a new or extended bathroom above an existing one.

Master Bath

The typical master bathroom is one that serves the master bedroom, offering more amenities—and some duplications—to serve the needs of two users. For example, a master bathroom may include a bathtub and separate shower as well as a lavatory or vanity with two sinks. Some master bathrooms have a separate space for the toilet or even two toilets.

Here are some suggestions for planning the installation of your remodeled master bathroom:

◆ If you install a skylight or solar tube (a smaller, round skylight in a tube) in your master bathroom, make sure it has protection against excessive room heat during the hottest part of the day.

◆ Upgrading the flooring or wall paint can dramatically increase the beauty of your master bathroom at relatively low cost.

◆ If you have selected a 72 in. cabinet for your vanity, make sure you can get the solid one-piece top to the room for installation. They aren't just big, they can be heavy.

◆ If you are doing your own labor, plan 100 to 150 hours for a complete master bathroom remodel and adjust according to your experience and how much uninterrupted time you can spend on the project.

◆ Be sure to gather up all tools, materials, and components before starting the job.

If you need to do some rough plumbing—moving pipes, flanges, or fixtures—refer to Chapter 13. If you need to do some wiring, install heating and ventilation, or install special cabinets, refer to the other advanced remodeling chapters in Part 3.

Garden Bath

A garden bath features plants, waterfalls, and other outdoor elements. In fact, many garden bathrooms are built with a sliding door or adjacent sunroom that allows occupants to step into a protected—and private—space.

Besides the common tub/shower/toilet/lavatory combination the garden bath may have a rock (or fake rock) waterfall in the corner, a door to a sunroom, an extensive area for plants, or other natural looking components. Each of these components will require additional planning, work, and cost. They are included in your bathroom remodel design (see Chapter 5). However, there are tips from the pros on making sure that you get what you think you ordered:

◆ Make sure that your plans include new *GFCI* electrical circuits for powering pumps and lights for any waterfalls.

Bathroom Words

GFCI stands for **ground fault circuit interrupter,** an electrical safety device that senses any shock hazard and shuts off a circuit or receptacle. It's typically required in any receptacle that is near a water source.

- If you are using craftspeople for special jobs (tile, terrazzo, rock work), make sure you coordinate other work so that everything is ready for them when they arrive.

- Even if your plans call for a specific cabinet, you typically can upgrade it before installation as long as you don't change its dimensions.

- If you have a question about whether a replacement component meets or exceeds building code requirements, call the building department that issued the building permit.

Potty Training

If you have open space in your garage, consider using one bay as a staging area for the installation of your new extended bathroom. Gather major fixtures as well as any plumbing, electrical, framing, and other components you'll need. Keeping everything in one place—and out of the way—can make the job go smoother.

Spa Bath

A spa is a mineral springs resort, named for the town of Spa in Belgium. However, *spa* has come to mean a fixture or a room that includes a jetted bathtub (sometimes called a "whirlpool" bathtub) and other amenities for relaxation and therapy. Here are some tips on planning your spa bathroom.

- Jetted tubs are heavy and bulky, so plan on having extra help when bringing the unit into the bathroom being remodeled.

- Seriously consider having a contractor do the actual movement and installation of a jetted bathtub to save injury to yourself or the expensive equipment.

- Make sure you protect any installed flooring and cabinets from damage when moving or installing the jetted bathtub.

- Verify that the subfloor into which the jetted tub will sit can safely hold the weight of the unit with maximum capacity of people and water.

You'll learn more about selecting and installing jetted bathtubs a bit later in this chapter.

Plumbing an Extended Bath

When it comes to extended baths, your imagination and your budget are the only limits. Extended baths can be homes unto themselves with a relaxing jetted tub, waterfall, garden, sunroom, large lavatories, lots of storage, mood lighting, and even a TV-music entertainment center and refrigerator.

That means selecting components and fixtures can be daunting. Actually, if you have this much space and money you're probably opting for a professional decorator who can help you narrow the choices to a reasonable list. Following are some guidelines.

Selecting Jetted Tub Baths

A jetted tub is an aerated bathtub. That is, air is entrained in the warm water that is circulated within the tub. The bubbles and the water streams have a soothing effect on muscles. The water streams, passed by jets, can be adjusted for direction and intensity.

Jetted bathtubs are available in a wide variety of sizes, shapes, power levels, and features. For example, you can buy a jetted bathtub that will exactly fit where you removed that 5 ft. bathtub. Or you can get a jetted bathtub up to 46 in. × 56 in. in size. Actually, you can buy larger jetted bathtubs, but they are then considered spas built for more than one.

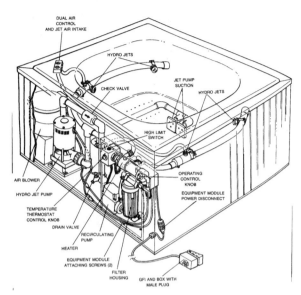

Components of a typical jetted bathtub or spa.

Jetted bathtubs larger than the standard bathtub are intended to sit inside a frame built on-site. Once the frame is constructed and the new plumbing installed, the jetted bathtub is lowered into place and connected. When completely installed, the frame is covered, typically with ceramic tile.

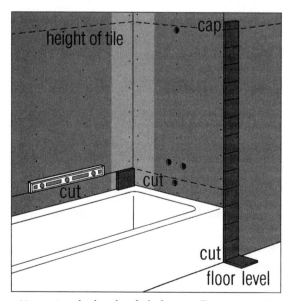

Measuring the height of tile for installation around a jetted bathtub.

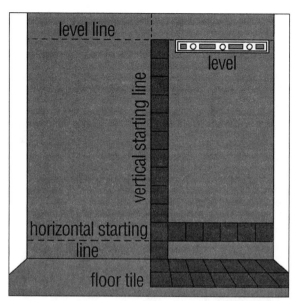

Make sure that your guide lines for installing tile are perfectly level.

Ceramic tile can be installed around and even above your new shower.

Jetted bathtubs can be recessed into the floor, but this requires structural remodeling. More homeowners opt to install the jetted bathtub at standard 12 in. to 14 in. rim height or to raise the tub on a platform.

Heads Up!

If the rim of the jetted bathtub is higher than 16 in. off the floor, make sure you include a step for safely getting in and out of the tub.

Selecting Special Showers

Special showers is a wide category covering any type of flowing water fixture other than the single standard shower head. They often have multiple adjustable shower heads at various levels including a hand-held shower. Some disperse horizontal sheets of water rather than cylindrical heads. Others have a shower bar made up of numerous smaller shower heads.

Because there are so many variations to the standard, it's best to shop around with a knowledgeable advisor, such as a bathroom remodeling designer or contractor. Alternately, you can educate yourself on the latest gadgets by investigating various bathroom shops. Even if your plans are already drawn and prep work is done there may be new fixture types that appeal to you and will require few or no changes in your remodeling plan.

Special tub/shower combinations require additional planning and framing for installation.

Selecting Saunas

A sauna is a room where the bather is cleaned with steam rather than water (the steam opens up the pores to remove impurities). Typically, water is poured over red-hot rocks to produce the steam. Many people prefer a sauna over an immersion bath or a shower. Others want all of the above for their bathroom.

Because a sauna is a room, construction is covered more fully in Chapter 16 on walls and flooring systems. But I'll give you an overview here of the process for those who don't need advanced instructions.

Once the room is built, wired, and plumbed, the sauna heater unit and controls are installed. Make sure that the room you build is air-tight so that moisture doesn't seep into walls, floors, or other rooms. In addition, make sure that the sauna heating unit you select is adequate for the room you build. If underpowered you won't get adequate steam and if overpowered you could get too much. It's better to get one a little larger than needed and turn down the controls as necessary.

Installing Extended Bath Fixtures

It's difficult to guess the amount of time you'll need to install all the fixtures for your extended bathroom. It depends on what you are putting in, how elaborate your design is, how much you plan to do yourself, and how handy you are.

An extensive spa bathroom with hot tub and waterfall shower or a sauna can take a remodeling professional 300 hours to install, excluding preparation work. If you are an experienced do-it-yourselfer, add up to 100 hours. If this is your first major project, double the time requirements—and seriously consider hiring a pro plumber for the rough-in or inside-the-wall work.

The following information will guide you in planning the installation of your extended bathroom as well as help you determine how much of it you want to tackle yourself. Let's first consider the tools and materials you'll need.

Tools and Materials

The tools you'll need depend on the complexity of the jobs you tackle. That makes sense. For example, installing a jetted bathtub replacement for your standard bathtub will take little more than what installing a tub takes (see Chapter 11). Installing a raised jetted bathtub will require structural framing tools (circular saw, power drill, etc.) as well as tile cutting and installation tools.

Potty Training

As you plan your bathroom remodel, don't forget to plan for power. Make sure you have a source of electrical power and heavy-duty extension cords as needed to give life to your power tools and work lights!

Installing a sauna requires construction tools such as saws, hammers, and wrenches. You're building a room. In addition, you may need a vapor barrier between the wall surface and the wall framing to keep moisture from seeping out. Refer to your remodeling plan's materials list for specific materials you will need to build a sauna room.

Preparation

Preparation for installing a jetted bathtub, spa, or sauna can be extensive. You need to make sure that not only are all of the fixtures and materials to be replaced out of the way, you also need to install all rough plumbing and wiring. This

may mean structural changes or additional services. Prep can take as long as installation.

Jetted baths that are built in place with frames require waterproof underlayment.

Shower tiles require underlayment.

Fortunately, with a good remodeling plan and the manufacturer's prep instructions you should be able to do the needed preparation without problems. If in doubt, ask the supplier of your primary fixtures for assistance. Some offer extensive written instructions on preparation while others have a telephone help line that you can call.

Installing Jetted Bathtub Baths

Replacing a standard bathtub with a jetted bathtub is approximately the same as installing a tub (Chapter 11). The differences are that you may need to move some of the plumbing (Chapter 13) and add electricity for the jet pump (Chapter 14). If you are installing a jetted bathtub that doesn't have finished sides you'll need to approach the installation differently.

1. Mark out the location of the jetted bathtub and determine where the water lines and drain are to be located, then install them (Chapter 13).

Install waterproof drywall, called greenboard, around the perimeter of your tub or shower before installing the finish surface.

2. Determine the location of the pump and follow manufacturer's instructions for bringing electricity to the pump, heater, and other electrical connections for the jetted bathtub.

3. Place the jetted bathtub in position and make sure it is level.

4. Make all plumbing and electrical connections according to the manufacturer's instructions.

5. If the sides of the jetted bathtub are unfinished, build and cover a frame around the tub using waterproof backerboard. Make sure you include a door for accessing the pump and electrical connection. Finish the sides with tile or other decorative surface.

Installing a fully tiled bathtub requires dense backerboard to give a firm foundation to the tile.

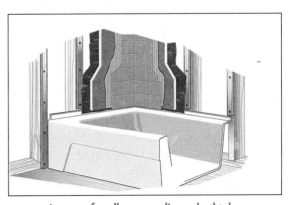

Layers of wall surrounding a bathtub.

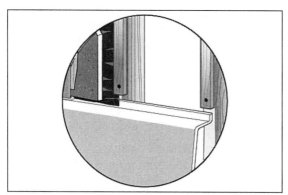

The bathtub's rim should be installed under surrounding walls so water doesn't leak around the edges.

Potty Training

If the jetted bathtub you are installing has a full base that surrounds the entire tub, it's easier to construct the base first, then install the tub.

6. Use waterproof caulk to seal all seams around the perimeter of the tub and base.

Note that local building code will probably require a GFCI receptacle for the jetted bathtub because it is close to water. GFCIs have a built-in circuit breaker that trips if there are any electrical grounding problems. To reset the receptacle you must be able to get to it. Therefore you will need an access door if you enclose the receptacle. Alternately, you can install the GFCI outside of the jetted bathtub base, but the cord may be unsightly.

Installing Special Showers

Installing a special shower is similar to a standard shower. The primary difference is in the shower heads and controls. Waterfall showers will require special heads built in or on the wall. Shower bars typically don't require special construction of the shower stall.

The best advice is to follow the manufacturer's instructions for installation of special shower components, or hire an experienced plumber to do the job. If you're a do-it-yourselfer, consider hiring a plumber to oversee your work including building and preparing the stall.

Installing Saunas

Saunas can be as simple as an ex-closet retrofitted as a one-person sauna with a heater or it can be a larger room with a heater and built-in shower. Most saunas are installed indoors adjacent to a bathroom, but they also can be installed outdoors as a separate structure.

Saunas can be planned from scratch, customized to fit your home, or they can be purchased as indoor kits or outdoor prefabs. Kits require your labor, but typically come with good assembly instructions. Make sure the manufacturer you select also has a customer service line that you can call if you run into problems.

Assuming that you are building your sauna from scratch, here are the steps to sauna construction:

1. Construct the walls either as tip-ups or built-in-place (Chapter 16) following your building plans. Make sure you include vents according to the manufacturer's instructions.

2. Install electrical receptacles and switches for the heater units and lighting.

3. Insulate the room per the building plans, then install the vapor barrier.

4. Install the flooring material.

5. Install paneling on the ceiling and walls.

6. Install trim molding covering joints between flooring, walls, and ceiling.

7. Install wall supports and slat benches.

8. Install the door jamb, pre-hung door, and trim.

9. Install the heater unit, control panel, and lighting fixtures.

10. Test your new sauna room.

Of course, there's more to remodeling an extended bathroom than these short instructional lists. Some do-it-yourselfers can build from them, but most need additional step-by-step instructions on specific aspects: plumbing, wiring, heating, walls, cabinets, and fixtures. Part 3 of this book offers clear instructions that you can use when remodeling any bathroom.

The Least You Need to Know

- ◆ Extended bathrooms require more work to remodel, but offer additional living and relaxation space.

- ◆ The simplest jetted bathtub to install is one that is of standard bathtub size and design and doesn't require additional framing.

- ◆ If you're not comfortable installing a new extended bathroom, hire expert help for the difficult parts.

- ◆ To make the most of your remodeling time, gather the tools and materials needed and have them standing by when you start the job.

In This Part

Part 3

Remodeling Bathroom Systems

Some bathrooms require more work than others. If *your* bathroom needs extensive plumbing changes, new electrical circuits, a heating and vent system, new walls or floors, or other major changes the instructions in Part 2 just weren't enough to get the job done. You need to know more!

This part offers *more!* Besides additional step-by-step illustrated instructions it gives you a deeper look into the primary systems that make up your bathroom. You'll see more of how they work, how to select and install them, and how to do it safely.

Best of all, you'll get tips from numerous professional remodelers on how *they* save time, money, and injury. With our help you'll soon be able to brag, "I remodeled it myself!"

In This Chapter

◆ Understanding basement plumbing systems

◆ Selecting tools and materials for the job

◆ Following local plumbing codes

◆ Remodeling plumbing systems

  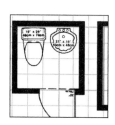  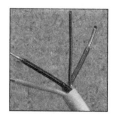 

Plumbing Systems

Bathrooms require water and water requires a method of distribution, called plumbing. Doing most any remodeling job in a bathroom requires some knowledge of plumbing. Even if you're just replacing faucets, you need to know how they fit into the scheme of things and how to make sure your job doesn't become a bigger job.

This chapter takes you inside the typical home's plumbing system to show you how it works—and how to remodel it. Even if you already have some experience with plumbing you'll learn some new tips and techniques. If you're new to plumbing, it will give you a basis for tackling your bathroom remodel project—or help you decide whether to call in a professional plumber. You'll discover that plumbing systems are quite logical and efficient. They're also relatively easy to remodel.

How Plumbing Works

Plumbing is your home's water distribution and waste collection system. Fortunately, it's actually *two* interdependent but not connected systems.

One system delivers cold and hot water to each of the fixtures in the room, as appropriate. All of the fixtures in the bathroom except the toilet require hot water. (Most bidets require hot water.) Because you don't want cold and hot water to commingle until it gets to the faucet, the water distribution system has two separate subsystems: hot and cold.

The other system removes drain water from sinks and waste from toilets, meantime venting gases and odors. The waste and gases are all taken outside of the home for distribution.

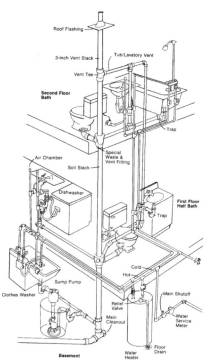

Components and layout of the typical residential plumbing system.

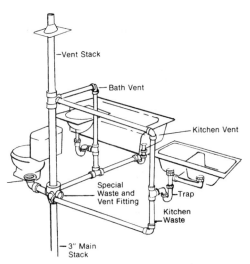

Components and layout of a typical bathroom and kitchen plumbing system.

Vertical pipes vent gases and odors out the top of the house.

Plumbing pipes run in the wall and under the floor of the typical new bathroom.

Your home already is constructed so making changes to the plumbing system requires remodeling it. How extensive the remodel job is depends on what you're trying to accomplish. Are you moving primary fixtures or simply replacing them? Let's take a closer look at how the plumbing system and its subsystems work together to distribute water and collect waste.

Delivering Water

Your home's water distribution system begins where the water supply enters the house. For homes supplied by municipal water systems the main water line enters the home near ground level on the side of the residence closest to the street or the main distribution line (alley, common line). For homes on private wells or springs the line typically enters the house on the side that's closest to the source.

Municipal water supplies are metered so they know how much to charge you each month. The meter may be located where the main water line enters your property or it may be installed inside your home, depending on local preferences and weather conditions. In this water line is the main shutoff valve that can stop flow of water to your home. You should know where this main shutoff valve is as you begin working on remodeling your home's plumbing system.

The cold water pipe system distributes water from the main water pipe to various areas and fixtures around the home: bathtubs, toilets, lavatories, kitchen sink, clothes washer, etc. It also delivers water to the hot water heater and system that, in turn, distributes heated water to various fixtures.

In older homes, cold and hot water is distributed in galvanized steel pipes assembled with threaded connectors. Installing or remodeling these systems often requires an experienced plumber, though you can buy *fittings* that will let you retrofit steel pipe systems with modern materials with little effort.

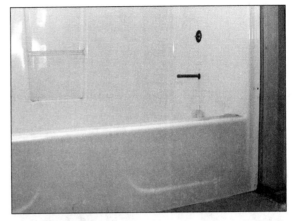

Plumbing pipes protrude through the shower wall and are finished by installing faucets and other fixtures.

Today's homes typically use copper pipes for new installation, though some local building codes allow for use of polyvinyl chloride (PVC) plastic pipes. Water supply lines typically are ¾ in. in diameter for trunk (main) lines and ½ in. for *branch* lines. The primary water line coming into the home could be 1 in. or larger.

> **Bathroom Words**
>
> **Fittings** are any devices that connect pipe to pipe or pipe to fixtures. A **branch** in a plumbing or heating system is any part of the supply pipes connected to a fixture.

In addition, various branches of the water distribution system may include a shutoff valve, allowing you to turn off hot and/or cold water to a specific room during remodeling without depriving the rest of the home of water.

Removing Waste

The waste system takes soapy water from sinks and waste water from toilets for removal from the home. Because the waste pipe system accepts drain and waste water as well as venting gases it's called drain-waste-vent or DWV.

Modern DWV pipes are made of plastic and easily installed in walls.

This DWV line includes a trap inside the wall rather than under the fixture.

The DWV pipes in older homes are made of cast iron. Fortunately, adaptation parts are available so it's easy for the do-it-yourselfer to install or remodel. Newer building codes allow for the use of plastic drain pipe, a relatively easy job for homeowners. Check with local building codes before installing new pipes. In fact, some building codes will require that you replace pipes in existing bathrooms you are

remodeling with newer materials. Parts are available at plumbing stores to help you add plastic drain pipes to a cast iron system.

Tub/shower plumbing includes copper hot and cold lines to the faucets and PVC drain lines.

The larger line is the drain and the smaller lines are hot and cold water for a bathroom lavatory.

Waste pipes are 1½, 2, 3, or 4 in. in diameter depending on local code and how much waste is expected to run through them—and how fast. For example, the main drain from your home to the municipal sewer line or your septic system typically is 4 in. in diameter. A toilet line may be 3 in. and a sink line may be 2 in., all depending on traffic and code. Drain pipes can be as small as 1½ in. for some applications.

A typical toilet drain line is 3 in. in diameter and located 12 in. from the nearest wall.

Waste lines include a *trap*, a curved portion of the pipe located near the fixture. Water sitting in the trap serves as a seal between the atmosphere in the waste pipe (yuk!) and the room.

Bathroom Words _____

A **trap** is a U-shaped drain fitting that remains full of water to prevent the entry of air and sewer gas into the building.

Where is the trap for a toilet waste line? It's inside the toilet, the S-curved portion under the water bowl. That means when you remove a toilet there is no trap between the room atmosphere and the waste line. When installing or replacing a toilet, plumbers stuff a wet cloth in the toilet flange to keep waste gases from entering the room.

One more point to make about waste pipes: To be effective they must have a minimum of ¼-in.-per-foot (2 percent) minimum slope to them so waste will move toward the sewer or septic system. If the slope is much more than this the water will drain too fast to move the waste; if the slope isn't sufficient all waste will sit where it is. Neither option is a good one. The trickiest part of installing or retrofitting a waste system is getting the slope right. Local building codes and your building inspector will give you the specific angles.

Potty Training _____

Which pipe is which? When digging around behind a fixture or in a wall you come across a pipe; how can you tell whether it is supply or drain? Water supply pipes are 1 in. (thickness of a large thumb) or less in diameter and drain lines are larger. Hot or cold supply line? Touch it.

Venting

As noted, waste line traps keep sewer gases from entering the rooms. Where do these gases go? Modern plumbing systems include a system of vent pipes that allow the gases to escape to the atmosphere *outside* of the home. Water is heavier than air so it makes sense to have these vent pipes exit out the top of the residence. They are single pipes with the end open but covered to keep rain out. Most homes have more than one vent pipe, depending on how close together the plumbing systems inside the home are located. It may be more efficient to have separate vents rather than to tie them all together.

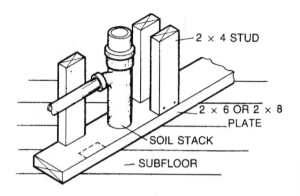

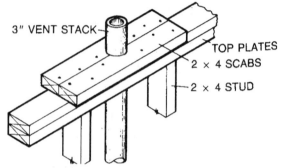

Details of how vent pipes pass through walls.

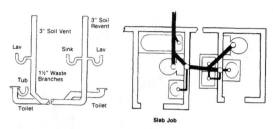

Installation details of a bathroom built over a concrete slab.

A *soil stack* is the main vertical pipe into which all waste from the home is delivered for distribution to the drain line. A *vent stack* is the main vertical pipe that vents gases from the soil stack. Technically, any pipe that carries waste is called a soil pipe.

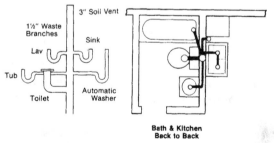

Installation details of a bathroom that shares plumbing with an adjacent kitchen.

Selecting Supplies and Tools

If you're planning to do your own plumbing, you'll need some materials, supplies, and various tools. Which ones? How can you be sure you select the right ones for the job?

Selecting materials may be made easier by your local building department. They probably will require a detailed remodeling plan for your project including what type and size of pipes you'll be installing. In fact, they probably won't issue you a building permit without knowing that what you install will conform to local building codes. So you or your designer will have to specify the materials you'll install.

Fortunately, building codes are relatively standardized so you won't have to guess too much. They may tell you that "copper is required for all new water lines and PVC is okay for waste lines, but make sure that you use the correct couplers to the existing system." That means you will have to identify the type and location of connection points in the existing system where you can install a new branch, then find plumbing couplers that will match up the old and new systems. Plumbing supply stores can help you select the correct components.

Materials

The materials from which you will build or remodel your plumbing system include plastics, copper, steel, and other components. Here is an overview:

- Galvanized steel pipe was previously the standard material for household plumbing to distribute supply water. It is primarily used today for repairs to existing systems.
- Cast iron pipe was previously used for drain and vent systems, now replaced by PVC in most applications.
- Polyvinyl chloride (PVC) plastic pipe is the preferred material for drain and vent installations. It typically is white in color.

PVC pipe includes markings to indicate size and material.

- Acrylonitrile butadiene styrene (ABS) is similar to PVC pipe, but not as rigid. It typically is black in color.
- Rigid copper pipe is the primary supply pipe material today. It is joined with soldered (sweat) fittings.

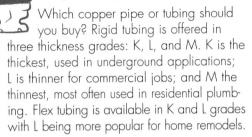

Potty Training

Which copper pipe or tubing should you buy? Rigid tubing is offered in three thickness grades: K, L, and M. K is the thickest, used in underground applications; L is thinner for commercial jobs; and M the thinnest, most often used in residential plumbing. Flex tubing is available in K and L grades with L being more popular for home remodels.

- Chrome-ribbed copper pipe is used as fixture supply tubing such as between a shutoff valve and a toilet.
- Chlorinated polyvinyl chloride (CPVC) plastic pipe is sometimes used for supply systems if allowed under local building codes. It typically is white in color.
- Cross-linked polyethylene (PEX) tubing is the latest plumbing material able to withstand extreme temperatures with flexibility and ease of installation.

Copper pipes are still the leading material for supply lines, but plastic pipes and PEX tubing are gaining. In addition, plastic and PEX plumbing materials are easier to cut and to install, making them the best choice for do-it-yourselfers.

Bathroom Words

Some plumbing codes will require a **dielectric union**. It's not an ashram or a rock band. It's a fitting made of two materials such as brass and steel with a plastic bushing in the middle. It minimizes corrosion and is required in some local building codes. (Corrosion invites leaks!) Ask your local building department and/or plumbing materials supplier.

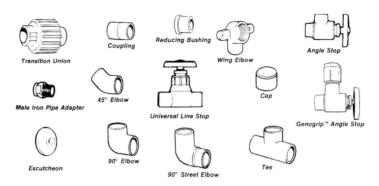

Various plastic plumbing fittings.

Supplies

Supplies needed to perform plumbing remodels and repairs depend on the pipe material used. For example, copper pipes require copper joints and fittings while plastic pipes use plastic joints and fittings. Copper joints can be made by soldering the pipe and fitting together. Fittings can be attached by friction and compression using compression rings.

Copper and plastic fittings come in a wide variety of structures and functions. A T-fitting, for example, allows one pipe to feed into two. An elbow allows the pipe to turn a corner. A union joins two pipes. There are many variations to these components including fittings that allow you to attach a new plastic or copper line into an existing steel or iron pipe. In addition, you may need hangers and straps to fasten pipes firmly in place.

Add to your list of supplies any products you need to seal, such as tub-and-tile caulk, plumber's putty (for around drains), pipe joint compound, solder and flux, and PVC primer and cement. Also include Teflon pipe dope tape for wrapping pipe threads for a better seal.

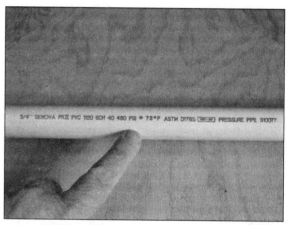

Household plumbing must meet specific pressure and temperature requirements. Make sure the pipe and fittings you purchase meet or exceed the minimum rating or "schedule."

Heads Up!

Make sure you use *lead-free plumbing solder* and not electrical solder that contains a small amount of lead. Otherwise the joints will soon leak and you'll have to do them again!

The best advice for selecting plumbing supplies is to sketch out what you're trying to do, then take the sketch to a plumbing supply store. Knowledgeable clerks can help you select the components you'll need and even give you some tips on how to install them.

Plumbing Tools

Tools needed for plumbing include the basics as well as some specialized tools. The basics include common screwdrivers, wrenches, pliers, and grips that are needed for fixing most things mechanical. You also may need a butane torch for copper pipe installation. Specialized tools include pipe wrenches, spud wrenches, chain wrenches, and faucet-seat wrenches.

In addition, some pipe materials require special tools for cutting and fitting. For example, copper pipe is cut with a tubing cutter, the ends are flared with a flaring tool, and the pipe can be bent with a tubing bender. PVC pipe is cut with a hacksaw; joints are fastened using special primer and adhesive.

You also may need some woodworking tools for your plumbing jobs. The tools include hammers, tape measures, and drills.

Health and Safety Code Requirements

You wouldn't want to buy a home that was built without any plumbing codes. Inside the wall would be the cheapest plastic tubing. The waste water would be stagnant and without a vent. The toilet may not have a water line running to it. You get the picture: Plumbing codes are for your benefit.

Yes, some of the codes are antiquated and seemingly intended only to keep plumbers in business. However, more municipalities are making plumbing and other building codes more homeowner friendly and even offering assistance in planning bathroom remodels that meet code.

So once you have an approved building permit—or even before—talk with local building inspectors about the local plumbing codes. How can you get a printed copy of the local codes? Are there classes available through the local college or trade school? Because most plumbing codes are standardized you can go directly to the source for copies of the code. Contact:

> National Standard Plumbing Code
>
> National Association of Plumbing, Heating, and Cooling Contractors (NAPHCC)
>
> P.O. Box 6808
>
> Falls Church, VA 22046
>
> Phone: 703-237-8100
>
> Website: www.naphcc.org

The plumbing codes are written for professional plumbers rather than do-it-yourselfers, so you may want one of the consumer books on this topic. Check Appendix B for additional information.

Advanced Plumbing

Chapters 10, 11, and 12 included basic instructions for installing bathtubs, showers, toilets, and lavatories. But how do you get the plumbing to the point where all that's left are these fixtures?

This chapter on plumbing continues with how to work with common supply and waste pipe. The most common materials are copper, plastic, iron, and steel. Here are step-by-step instructions for installing and remodeling plumbing systems. Your remodeling plans tell you "what" and the following show you "how."

The latest concept in plumbing is called *manifold plumbing*. Once the water line enters the house it enters a manifold with multiple outlets, each with its own shutoff valve. Each outlet has a single clear plastic tube that runs directly to a specific fixture such as a bathroom or kitchen sink, thus eliminating branch lines that serve multiple fixtures. Installation is not only easier than traditional branch plumbing, it is also easier to isolate a specific fixture and turn it off at the source for repair. National building codes allow manifold plumbing systems in many parts of the United States. Check local codes.

Rigid plastic, flexible plastic, and copper are all popular materials in modern plumbing.

Manifold plumbing splits the main water line into individual lines.

Each line serves a specific fixture such as a bathroom faucet or a toilet.

The flexible PEX lines terminate at a fixture.

Installing Copper Pipe

Copper pipe or tubing is much easier to install than galvanized steel, which copper replaced for water systems in most home construction about 50 years ago. Steel pipe was not only heavy it required special threading tools for connections. Steel pipe is still used for most natural gas and many other lines, but is only used to repair older water supply systems. For new construction and remodeling, copper is the popular choice—or requirement.

As mentioned earlier in this chapter, there are two methods of connecting copper pipe: soldering and fitting. Many systems use both, soldering all connections except those that are on or near fixtures. Check local plumbing code for requirements.

Here's how to cut copper pipe:

1. Measure and mark the pipe to length.

2. Use a copper pipe cutter to score and then cut the pipe. To cut, rotate the cutter around the pipe once, then slightly tighten the cutter and rotate it again, continuing until the pipe is cut through.

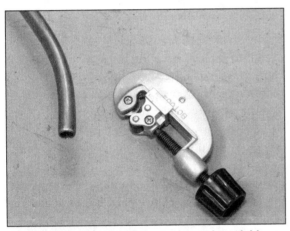

Copper pipe cutters are easy to use and available at larger hardware and home repair centers.

3. Use a burr remover or brush to remove excess metal from the new edge.

4. Use an emery cloth or pipe surface tool to clean the end of the pipe that will be soldered or coupled.

Here's how to solder copper pipe:

1. Coat the end of the pipe with flux (soldering paste) as specified by the manufacturer to remove oxidation.

2. Repeat the process for other pipes that will be connected at the fitting.

3. Insert the prepared pipe tips into the fitting, turning each one to evenly spread the flux.

4. Use a torch to heat the joint for a few seconds, then continue heating as you apply solder to the joint.

Heads Up!

Don't heat the solder, just the pipe. Placing the solder against the pipe will make it melt or "sweat" and flow into the joint to fill it. That's why they are called "sweat joints."

5. Carefully clean the joint with a rag and check for leaks. Reheat and add solder as needed for a leak-proof joint.

Here's how to fasten copper pipe fittings:

1. Cut the pipe to the appropriate length (see above).

2. Slide a coupling nut and then a compression ring over the end of the pipe.

3. Align the fitting with the end of the pipe and move the compression ring near the fitting.

4. Coat the compression ring and the fitting with pipe joint compound.

5. Connect the coupling nut to the fitting and carefully tighten.

Installing Plastic Pipe

Plastic pipe is easier to work with than copper pipe because connections are easier to make. Here's how:

1. Measure and mark the plastic pipe to length.

2. Cut the pipe squarely using a miter box, hack saw, or plastic pipe cutter.

3. Clean the end of the pipe of debris with an emery cloth.

4. Check to make sure that the fitting snugly fits the pipe. If an elbow is to be set at a specific angle, mark both the pipe and fitting with a pencil to indicate correct position.

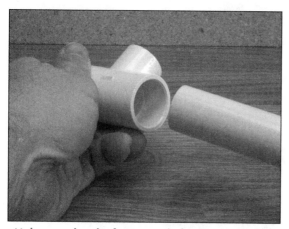

Make sure that the fitting snugly fits the pipe before cementing the two together.

5. Apply plastic pipe primer to the fitting as well as to the area of the pipe where the fitting will be installed. Note: Only use a primer recommended for the type of pipe you are installing.

6. Apply the appropriate type of plastic pipe cement to the fitting and the end of the pipe.

7. Assemble the pipe(s) and fitting, making sure you match up any alignment marks (step 4).

8. Firmly hold the assembly for 10 to 15 seconds while the cement begins to adhere.

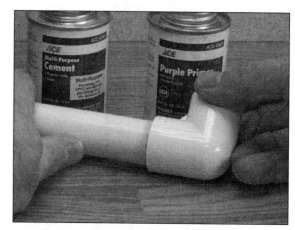

Apply the primer, then the cement.

Retrofitting Steel Pipe

Again, steel pipes usually aren't used today except to replace a section for repair. In such cases, purchase a precut and prethreaded length of pipe. Some plumbing shops will cut and thread a steel pipe to length.

You can retrofit steel pipes with special couplings that will fit between old steel pipes and new copper or plastic pipes. For these, make a drawing of your existing plumbing along with your new plumbing and take it to a plumber or plumbing supply store. Most such couplings are banded couplings similar to those on an automotive radiator hose.

Retrofitting Iron Pipe

Retrofitting an iron drain pipe with new plastic pipe can be challenging. The connection isn't that difficult to make—similar to the coupling for a steel pipe. The hard part is getting to the pipe itself. In some cases it is buried in dirt near or under the house. In the worst cases it is located under a concrete slab in a basement or under a patio. Either situation requires some searching and digging.

You sometimes can find the path of an unseen cast iron pipe by using a metal detector and some logic. If you know where it enters the house or the room and where it probably is a metal detector can help you determine its exact location—much easier than tearing up an entire concrete floor.

Once your plumbing system is remodeled you can continue on with other systems (wiring, heating, walls, and cabinets; see Chapters 14 through 17) or you can install plumbing fixtures (see Chapter 18).

The Least You Need to Know

◆ Plumbing is a water distribution and waste collection system that must conform to health requirements and building codes.

◆ Plumbing components are standardized so you can select them yourself at any plumbing supplies retailer or larger building materials store.

◆ Copper and plastic pipe systems make remodeling bathrooms much easier for the do-it-yourselfer.

◆ Even though iron and steel pipes are no longer used for most residential plumbing systems you can retrofit them with new copper and plastic pipes using couplers.

In This Chapter

- ◆ Understanding electricity and lighting systems
- ◆ Choosing electrical materials and tools
- ◆ Following electrical code for remodelers
- ◆ Working safely with electricity
- ◆ Doing your own wiring

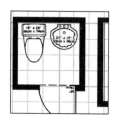

Chapter 14

Wiring and Lighting Systems

Bathroom lighting is often overlooked when remodeling. Yet bathroom lighting systems can dramatically enhance the function and beauty of any bathroom. Especially if natural light isn't readily available, well-placed artificial lighting can change the room's atmosphere more than just about any other remodeling component. And it doesn't hurt to have more outlets and switches, either.

In this chapter, you'll learn how electrical and lighting systems are planned and installed in bathrooms. Even if you've never taken the cover off a light switch you'll find sufficient instruction and step-by-step photos here to tackle most electrical remodeling job—or at least keep your electrician honest.

How Electricity Works

Electricity is the flow of energy. Thick books have been written about the science of electricity and how it works—not important. What is important is that it *does* work to produce light, operate appliances, and do other jobs around the home.

Your home is an existing structure. Your new or remodeled bathroom might or might not already have sufficient electrical service for the jobs you intend. So you might have to extend your electrical system. Or you might only need to replace some of the circuit components— lights, receptacles, switches, and so on—that are already wired into place. In either case you need to know a little more about electricity so you can work safely.

The Electrical Circuit

Think of electricity as an oval track called a circuit. If something blocks the track in any location the circuit isn't complete and power can't pass. If a switch is *off*, a light burned out, or a wire disconnected the circuit is blocked and electricity doesn't flow. It's about that simple in operation.

Within that circuit are various components that require electricity to operate. Most are either wired in (lighting) or access the circuit's power through a receptacle (hair dryer). Your—or your electrician's—job is to analyze and, if needed, extend the existing electrical system in your home to do the jobs needed in the new or remodeled bathroom.

The first step in this process is to determine what electrical power is available in your home right now. The place to start is at the service panel, the point where electricity is distributed to the various house circuits. Electrical service comes into your home as 240-volt electricity with two hot (electrified) wires and a neutral wire. Electrical ranges and dryers use both hot wires for 240v service. Just about everything else in your home uses 120v service taken from *one* of the hot wires and the neutral (sometimes called the *ground*) wire.

The service panel is the starting point for each of the wiring circuits in your home. And each circuit has a safety switch that turns off the circuit if there is a problem. The safety switches are called circuit breakers or fuses, the weakest link in the circuit. If something goes wrong, the circuit breaker trips and must be reset or a fuse blows and must be replaced.

Each of the circuits has a maximum electrical rating, measured in amperes or amps, the measurement for electrical current. For example, a bathroom circuit may have a circuit with a rating of 15 or 20 amps (15A or 20A). If appliances and lighting on the circuit attempt to draw more current than that, the entire circuit is shut down by the breaker or fuse.

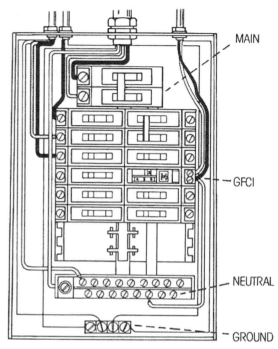

Inside details of a typical electrical service panel.

Potty Training

How many amps of service should a circuit have? It depends on what fixtures and appliances you expect will be using the circuit. A hair dryer uses up to 10A, a ceiling fan uses just 5A, but a window air conditioner can take 15A or more. The best formula is to add up all the loads in watts (W) and add 20 percent for safety. Then divide the total by 120V. If the total is 1600W, add 320W for safety for a total of 1920W, then divide by 120 for a total of 16A. A 15A circuit won't be enough; install a 20A circuit wire and circuit breaker.

So the first step in remodeling electrical circuits is to find out what your home has and how you can add to it if needed. To determine what's in place, add up the values of all the circuit breakers or fuses in the service panel.

Older homes often have a limit of 100A and newer homes 200A, though larger homes can have electrical service systems of 300A or more. If your current panel has breakers totaling 160A and you only need 20A more for your new bathroom all you'll need to do is wire in a new circuit and add a new 20A circuit breaker to the panel.

If, however, your panel already has the maximum amps dedicated to circuits you may have to start the job by replacing the service panel with a larger one—typically a job for an experienced electrician. Alternately, if your home receives 200A service but the main panel only has room for 100A of circuits you can have a subpanel installed to take advantage of the additional electrical service.

Wiring

Wiring a new bathroom typically isn't very difficult, especially if you've removed the old stuff and have access to inside the walls. You can install new wiring to fixtures with basic tools.

Electrical wiring is built in to a cable, multiple insulated wires that are then wrapped in a sheathing. A common wire cable for residential service is *Type NM 12-2 G* meaning nonmetallic (plastic) sheathing around two insulated 12-gauge wires and a bare ground wire. The two insulated wires are black (hot) and white (neutral). Another popular wire size is 10-gauge.

Heads Up!

Remember that the *smaller* the gauge number the *larger* the wire and the more electrical current it can carry. That means 10-gauge wire is larger and carries more electricity than 12-gauge wire.

Wiring typically is "pulled" or installed from the fixture to the service panel. That is, you install the electrical lighting fixture box and switch or the receptacle boxes and then run the wire from the boxes to the panel through walls. Cables for lights and switches run higher in the wall and in the ceiling. Cables for receptacles run low in the walls.

Once the cables are in place, remove any slack and *carefully* attach the cable to framing studs with construction staples. Your building department will tell you where cable needs to be stapled to keep it from moving around in the wall. Just make sure you don't put a staple *through* the cable or wires—a definite fire hazard.

Sometimes running new wiring in an existing bathroom can be a challenge. Just remember that you can open up a wall, trace the old wiring down, then close up the wall with new drywall paneling or a drywall patch kit (see Chapter 16). Or you can use a "fish line," a special tool for running wires in walls through small holes. Alternately, you can install wiring on the outside of the walls using special metal runs available through electrical supply stores.

Lighting

Lights are simply bare wires that make it so difficult (but not impossible) for electricity to run through them that they glow during the effort. The bare wires are surrounded by gases to enhance the light for the least amount of electricity.

Lighting fixtures are the components that house the electrical lights. Fixtures include the mounting plate, wires, bulb receptacles, and sometimes switches. The electrical wiring and mounting box are installed in walls during what's called the rough-in and fixtures are installed during the finishing phase.

Installing lighting fixtures is relatively easy as most manufacturers include an instruction sheet that shows how the fixture is installed according to the National Electrical Code (NEC) or, in Canada, the Canadian Electrical Code (CEC). For simple lighting fixtures, the black wire on the fixture is attached to the black electrical circuit wire, white to white, and ground to a grounding screw in the electrical box. Lighting fixtures that include fans require more complex wiring, but packed instructions will show you how it's done. If in doubt, hire an electrician.

Controls

Electrical circuit controls include switches and receptacles. Switches allow lighting or other fixtures to be turned on and off. Receptacles (outlets) allow you to plug in or unplug appliances for electrical service.

Potty Training

How many electrical receptacles can you put on a circuit? Most local electrical codes allow up to eight. Remember that if they are installed near lavatories, bathtubs, or showers they must be GFCI receptacles.

Electrical receptacles and switches are relatively easy for do-it-yourselfers to install. That's because they are standardized. The electrical box was installed with the end of the cable exposed (or you are removing and replacing an existing receptacle or switch). You simply prepare the end of the wire cable and attach it to the new component. Instructions follow later in this chapter.

Selecting Materials and Tools

The things you need for installing wiring and lighting are easy to find. Just about any building material store and even the hardware

departments of discount stores sell wire cable rolls, lighting fixtures, receptacles, switches, wire nuts, and other components. And they're cheap. You can buy a replacement switch for about a dollar.

Materials

Electrical components you might need for your bathroom remodel include:

- Electrical multiwire cable (per local electrical code)

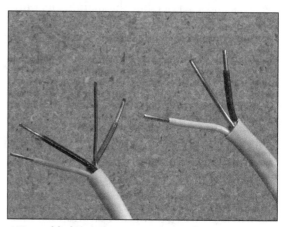

Wire cable has either two wires or three wires (plus a ground wire).

- Electrical boxes (metal or plastic, depending on application and local code)
- Electrical conduit or pipes to contain wires (required in some applications by local electrical codes)
- Electrical receptacles with screw terminals, push-in fittings, or both
- Electrical circuit switches (single pole, three-way, four-way, etc.)
- Electrical lighting appliances
- Solderless connectors (also known as wire nuts) for connecting wires together

Attach wires to electrical receptacles either at terminal screws or into holes in the back.

There are other components to some wiring jobs, but these are the most common. If your remodeling plan has a detailed wiring diagram it will include the location and type of the various components you need. In addition, a materials list for the job will include the specific electrical wiring and lighting components needed. It's your shopping list.

Electrical Tools

Tools for electrical remodeling are specialized, but most aren't very expensive. You can find them at larger hardware stores and building centers. Because of their relatively low cost, opt for better quality tools, especially if you expect to do much remodeling in the future. A good quality wire stripper can save time and effort for decades.

Common electrical tools for remodeling bathrooms include:

◆ Screwdrivers (Phillips and straight or standard)

◆ Pliers (needlenose and standard)

◆ Plug-in circuit tester (indicating condition of live circuits)

◆ Cable ripper (for removing plastic sheath from NM cable)

◆ Wire cutter and stripper (for removing insulation from wire tips)

◆ Circuit tester, continuity tester, or multimeter (for testing voltage, amperage, or resistance)

Electrical Code

Can you wire your bathroom's electrical system? Probably. Most local building codes allow homeowners to wire or rewire their homes. So the question is: *Should* you wire your bathroom? The answer depends on your experience with electricity, how much time you have, and your budget. Remember that many thousands of homeowners have successfully wired or rewired their homes. You can, too.

Most city and county building departments use a standard set of rules for electrical systems. In the United States it's the National Electrical Code (NEC) and in Canada the Canadian Electrical Code (CEC). Your local building department and technical bookstores have published resources that can help you understand and apply these electrical codes. Most methods make sense when you think about them.

Electrical codes standardize how electrical systems are installed for the safety of current and future occupants. For example, electrical codes typically require that receptacles (plugs, outlets) are installed 12 in. from the floor **and** no more than 12 ft. apart along walls. Code **also** dictates how many and what type of wires **are** used in residential electrical systems.

Heads Up!

Here's the bottom line: The purpose of all electrical codes is *safety*. Codes are written—and enforced—so that all future occupants of the building will be as safe from electrical injury as possible. So don't be offended if the building department's electrical inspector seems like a dictator. It's in the job description.

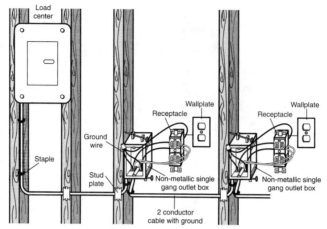

NM (nonmetallic) cable is run from the box to the electrical panel.

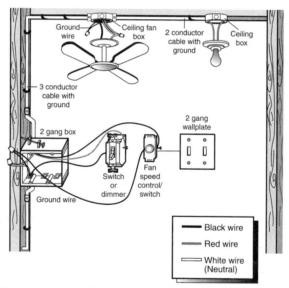

Fixtures are wired from and attached to the terminal box.

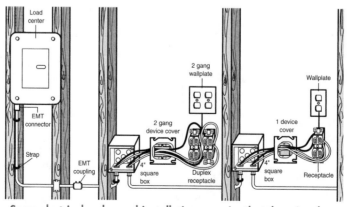

Some electrical codes and installations require that the wires be housed in metal conduit.

Advanced Electricity

Ready to remodel your bathroom's electrical system? Following are step-by-step instructions for the most common electrical jobs along with some clear photos to show you how it is done.

One important tip first: Work safely. Electricity *must* obey the laws of science! It will not shock you unless you give it a chance to. If you make sure that circuits are off before working on them, that you are not part of the conductive path, and that you use the correct tools and materials you will have no problems. However, if you forget or ignore the laws that electricity lives by you are opening up the opportunity for electrical shock. Don't be afraid of electricity; respect it and it will respect you.

Installing Electrical Boxes

As you've learned, electrical wiring starts in the service panel, an electrical box, and runs to the end of the circuit, another electrical box, where the fixtures or other components are installed. So the first step in setting up a new circuit is installing electrical boxes where they need to be.

Electrical boxes are attached to wall studs before the insulation and drywall are installed.

Fit the box snugly against the stud at the appropriate height (typically 12 in. above the floor) and fasten to the stud.

Electrical boxes, sometimes called terminal boxes, house the connection between the wires and the fixture or receptacle. Somewhere between the service panel and the terminal box there may be another electrical box or two. A junction box is one that protects an electrical splice such as where two or more electrical lines join up. A switch box also can be between the panel and the terminal box and house a switch that turns the circuit to the fixture on or off.

Potty Training

You'll learn a lot by being a complete idiot. Visit a large building material supplier's electrical department and tell the clerk "I need some help figuring out my first electrical job." Most clerks in this department are experienced electricians or advanced do-it-yourselfers who enjoy helping others. If you find one who isn't helpful, keep looking.

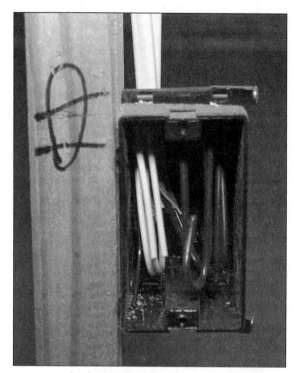

Simple electrical terminal box with wiring.

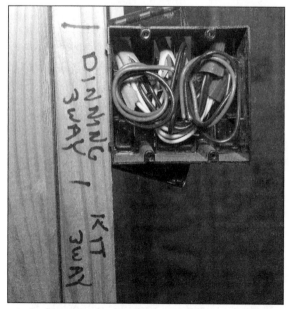

A single terminal can serve more than one switch or fixture.

Installing electrical boxes in new homes is relatively easy because the wall framing is in place but the drywall isn't. The electrician simply nails the boxes in the appropriate locations and runs wire cable from there back to the service panel. If your bathroom remodeling job requires that you open up or add new walls the process is the same for you. Boxes are mounted on studs or other wood components and wires are run through other wall members back to the panel.

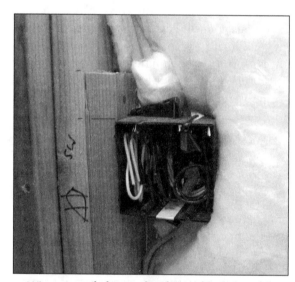

Wires are coiled up in the electrical box to make them easier to find once the drywall is installed.

However, what if you're not planning to tear up a wall covering or add a new wall? Then you can install *remodeler's boxes,* special electrical terminal boxes that can be installed through the drywall. Here's how:

1. Use an electronic studfinder to identify the location of the stud onto which you can attach the box.

2. Using the template that comes with the remodeler's box, cut a hole in the drywall for the box.

3. Remove the "knockout" hole in the box and pull the wires through.

4. Insert the remodeler's box through the hole and against the stud.

5. Nail or screw the box to the stud.

Note that some building codes require that you use metal boxes with locknuts that tightly grip the wire cable. Make sure you install the electrical boxes according to local code requirements—or be ready to remove and replace them if the building inspectors says so.

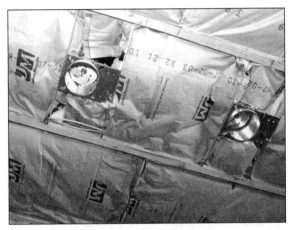

Recessed lighting is installed in special boxes and mounted on joists.

Installing Wiring

Once the electrical terminal box is installed in the bathroom, feed the wire cable from the service panel or a subpanel to the fixture. This may not be so easy depending on whether the wall is open or closed. If open you will drill holes in successive studs until you reach the fixture.

If the walls in between—or at least some of them—are closed, you will have to "fish" the cable through the walls. The easiest way typically is drilling a hole through a top or bottom plate in the wall above or below the electrical box, running the wire through the hole, then running it to the panel between floors of the home. Local electrical code will dictate more specifically what you can and cannot do. You

might decide to hire an electrician to do this job. Or you can buy a "fish line" at larger hardware stores and follow the instructions that come with it.

Waste Lines

When you run wire, remember to always allow an extra couple of feet of cable on each end. Cable is relatively cheap and you don't want to wind up with a short cable and have to replace it with a longer one.

Installing Receptacles, Switches, and Fixtures

Are you ready to finish the circuit? With a few basic tools you now can install electrical receptacles, lighting fixtures, and control switches in a relatively short time. Here's how:

1. Make sure that electrical power to the circuit is off.

2. Cut the wire cable to a length 6 to 8 in. beyond the front of the box.

Pull the wire from the box, cut it to length, and strip insulation from the end of the wires.

3. Use a cable ripper to remove the sheathing from the wire cable inside the box.

4. Use a wire stripper to remove the last inch of insulation from the end of each covered wire.

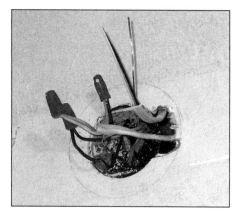

Wires are pulled, stripped, and capped with wire nuts.

5. Use a wire nut to connect the ground (bare or green) wire to the receptacle, fixture, switch, or box.

6. Use a wire nut to connect the white (neutral) wire to the receptacle, fixture, or switch.

7. Use a wire nut to connect the black (hot) wire to the receptacle, fixture, or switch.

8. Mount the receptacle, fixture, or switch on to the electrical box with screws (included) and install any cover plates.

Wired switch without the cover plate.

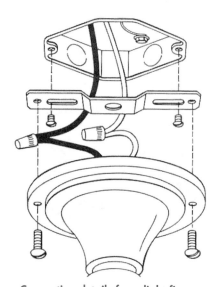

Connection details for a light fixture.

Some electrical components are a little more complicated, such as three-way switches and lighted fans, but most come with specific installation instructions.

Don't put away your electrical tools just yet. The next chapter covers installing and remodeling heating and ventilation systems for bathrooms.

The Least You Need to Know

- Electricity is a harnessed power system that obeys easy-to-learn laws of physics.

- If electrical remodeling is extensive, it may be faster and easier to remove all drywall first and replace it when done.

- Make sure the wiring plan for your remodel has sufficient detail and a materials list so you know exactly what you need and how to install it.

- Remodeler's boxes and other after-market materials can make remodeling a bathroom's electrical system much easier.

In This Chapter

- ◆ Understanding bathroom heating and ventilation systems

- ◆ Choosing the materials and tools needed for installation

- ◆ Installing bathroom heating systems

- ◆ Installing and replacing exhaust systems

  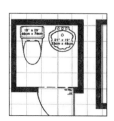  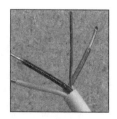 

Heating and Ventilation Systems

One of the primary reasons for remodeling any bathroom is comfort. Many homeowners choose a comfortable bathtub and other fixtures, but forget to consider two of the most important comfort systems: heating and ventilation.

Of course, once the job is done they realize that more attention should have been given to adequately heating or ventilating the room. Coming out of a shower, the floors are frigid and the mirror is fogged. And besides human comfort, walls and woodwork prefer a well-heated and ventilated bathroom. This chapter shows you how to make your bathroom's atmosphere more comfortable for everyone and everything.

How Heating and Ventilation Systems Work

The heating system has a simple job: warm the air that passes by. How that is done depends on many factors including the least expensive source of heating energy (electricity, gas, fossil fuel) for your location as well as the ease of getting the energy or heated air to the bathroom.

Bathroom ventilation systems, too, have it seemingly easy. All they need to do is remove stale air from the room. This can be done naturally, such as with an open window, or mechanically with a fan.

Then there are combinations. Bathroom exhaust fans are often combined with a light fixture, a fan-forced heater, or a radiant heat lamp. This device simplifies the selection and the installation of bathroom heating and vent systems. Let's take a closer look at heating and ventilation systems for remodeled bathrooms.

Heating Systems

Many remodeled bathrooms simply extend or upgrade an existing forced-air system with an additional duct to the room. Before extending a system make sure that your heating system can handle the extra requirements. Of course, if there was already a bathroom there and all you are doing is a facelift or a makeover the existing heat system may be adequate.

Heating ducts typically run through the ceiling or floor.

A suspended ceiling can hide HVAC ducting, plumbing, and electrical service from view.

Heating registers in the floor deliver warm air to the bathroom.

Because a bathroom is a place where you may not be clothed, consider supplemental heat for this room. It's a better option than turning up the heat throughout the house just to keep the chills away as you step from the shower. A small space heater installed in a wall near the shower or tub can not only keep you warmer, it also can help remove some of the moisture from the room. Just remember that many electrical space heaters require a dedicated 240v circuit. Others can be wired for either 120v or 240v power.

Potty Training

Heating systems are popular by region. Areas of the United States with cheap hydroelectric power prefer electrical heating systems. If oil fields are nearby, the preferred heating source may be fuel oil or natural gas. Coal is relatively cheap in other locations and serves as the primary heating fuel. If you are moving from one part of the country to another—as many do each year—make sure you learn what heating systems are preferred in the area and how to select and upgrade them.

Ventilation Systems

Bathrooms require either natural or mechanical ventilation to remove stale air and replace it with fresh air. You wouldn't want it any other way.

Natural ventilation includes an open window or other device that allows air flow to and from the home without also letting critters in. Of course, if you live in a climate where expensive air (heated and/or cooled) is required you don't want to lose it so a natural vent may not be the best option. If natural ventilation is an option, make sure the opening is at least 5 percent of the room's floor area. That is, a 100-sq.-ft. bathroom should have at least 5 sq. ft. of natural ventilation.

Mechanical ventilation is a fan. The fan should be located close to the shower and toilet and as far away from the entry door as possible. The ventilated air should exit outside of the home— not into a living space or attic—at least 3 ft. away from any opening into the home (window, door, vent intake).

Selecting Materials and Tools

Bathroom heating and ventilation systems can be cheap or expensive, easy to install or a logistic nightmare. Hopefully, the ones you've selected are relatively easy to install with basic tools—or you've decided to hire a pro to do the difficult work.

Let's consider the materials and tools you'll need for installing or upgrading one or both of these systems. Then you can decide how much of it you want to tackle yourself.

Materials

Bathroom exhaust fans are selected by their ability to move air a specific number of cubic feet per minute (cfm). Ideally, the cfm for your bathroom should be *eight times* the cubic feet of air in the room. To make the math easy, a 10 ft. × 10 ft. bathroom with an 8 ft. high ceiling holds 800 cu. ft. of air. Move that air eight times and

you have 6,400 cu. ft., the rating you would want for the exhaust fan in this bathroom.

Another consideration when buying an exhaust fan is the noise. Some, especially older ones that have worn components, can be quite noisy. Fortunately, exhaust fans are also rated for noise level, in sones, from 1 to 4. The lower the sone number the quieter the unit. As a comparison, a new refrigerator runs at about 1 or 2 sones in noise level.

Additional things you need for installing a heating and/or ventilation system in a remodeled bathroom include electrical supplies and some structural supplies. Electrical supplies include wire nuts, duct tape (for the exhaust duct), and maybe some electrical tape. Structural supplies include drywall screws or nails, silicone caulk, and roofing cement.

Of course, if you are extending or upgrading an existing forced-air system the materials you need include a new register and/or vent and some ducting. Heating ducts can be rigid aluminum rectangular tubes or flexible cylindrical tubes. The advantage to flex ducting is that it is easier for do-it-yourselfers to assemble and install.

Tools

The tools you need depend on the job you're doing. If you are retrofitting an old heater or exhaust fan with a new one you may only need basic tools: screwdrivers and wrenches. If you are installing a new unit where there wasn't previously one you will need a small saw (keyhole, saber, reciprocating) and a hammer at the least.

If you are installing a stand-alone gas heater you will need many of the tools and materials used for water systems such as pipe and plumbing tools. It's best to leave gas system installations to a professional. If you want to do it yourself make sure the appliances you purchase have thorough instructions and you review Chapter 13 on installing plumbing. Be aware that building codes have different material requirements for pipes that distribute gas than they do for those that distribute water.

Working Safely

Heating systems can be the most dangerous services in your home. Not only is the output hot but the processes that produce the heat can be dangerous. Electric furnaces and space heaters usually use 240v electricity. Gas furnaces and space heaters require natural gas, liquid petroleum, or propane gas. So make sure you read all instructions, especially the safety ones, that come with heating appliances.

In addition, cutting through drywall creates dust that can get into your eyes and mouth. The dust isn't toxic, but it is abrasive and can cause irritation or possibly damage.

Of course, working with any tool offers the potential of being hurt by it. Power tools can get you hurt faster than other tools, but they all can injure you.

The point is: Please work safely. Think about what you're doing, know the inherent dangers and avoid them. Thank you.

Potty Training

If you're working around dust or debris, make sure you wear a breathing mask and sealed safety goggles. Older homes, especially, can have old insulation and who-knows-what inside walls you're tearing in to. Make sure your eyes and breathing are protected.

Installing HVAC

HVAC stands for heating, ventilation, and air conditioning. Remodeled bathrooms require these services from the home. Fortunately, many homes already have adequate heating and air conditioning systems in place so the job is simply to extend or upgrade the existing system within the new or remodeled room.

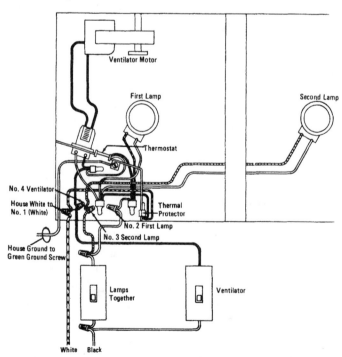

Wiring diagram for a typical exhaust fan and switch, similar to wiring for a space heater.

Bathrooms, though, benefit from supplementary air systems that heat or ventilate the room without putting additional burdens on the home's existing HVAC system. That's why the most common bathroom heating and vent jobs are to install or replace a space heater and to install or replace an exhaust fan.

Installing a Space Heater

If you are adding a space heater to your remodeled bathroom where there wasn't one, installation depends on whether you must cut a hole in the wall and run wiring or the drywall isn't yet in place and you can install the unit without obstruction. Let's assume the worst and see how a space heater is installed in a new location on a finished wall.

1. Use a stud-finder to identify the location of wall studs near your proposed site for the space heater. The location should be selected based on the shortest distance for wiring to be run.

2. Use a keyhole saw to cut a small hole in the wall near one stud to confirm the location of the stud.

3. Place the heater housing against the wall where it will be mounted and mark the perimeter on the wall with a pencil, then remove the housing.

4. Install the electrical circuit (see Chapter 14) to the hole where the heater will be installed.

5. Following manufacturer's instructions, attach the heater housing to the stud(s) and install the wall heater.

6. Install the cover plate on the wall heater, turn on the circuit, and test the heater.

Installing a radiant electric baseboard heater is similar to a space heater except that the heater is mounted on the floor and against the wall instead of recessed into the wall. In fact, the job is a little easier because the wiring can be run under the floor and into the unit rather than through the floor, then through the wall, and into the unit. The disadvantage to a baseboard heater is that most models don't include a fan to circulate the warmed air.

Replacing a Space Heater

If you are upgrading or replacing an existing wall-mounted space heater the job will be relatively easy if the units are identical in size and configuration. Make sure you get the model number and measurements of your existing system and take them when shopping for a new unit.

Electrical supply stores and larger home centers can tell you what replacement and upgrade models are available for specific model space heaters manufactured over the past few decades. In fact, the latest replacement of the old model may be quieter, more efficient, and even more powerful than what it is replacing. Even if you're not seriously thinking about replacing the existing space heater take a look at what is available.

Installing an Exhaust Fan

The most difficult aspect of installing an exhaust fan is determining where to place it for optimum ventilation. In most cases, remodelers install exhaust fans above and in front of the toilet, though it may be located between the toilet and the shower depending on the configuration.

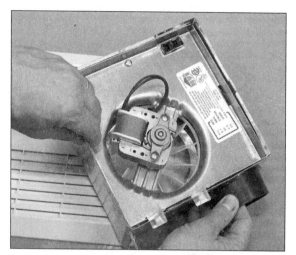

Typical exhaust fan system.

Air is pulled from the room by the fan and exhausted out the vent hose.

Once the probable location is known, here's how to install a standard exhaust fan with ductwork:

Potty Training

The job of installing a new exhaust fan is easier if you are thoroughly remodeling the bathroom and don't have drywall installed yet. You can install the fan box, duct, and even run the electrical wiring from inside the unfinished bathroom instead of the area above.

1. Drill a hole through the ceiling about where you will install the fan, then insert a wire through the hole so you can find the hole from above.

2. In the attic or crawl space above the bathroom, pull back the insulation where the wire has come through.

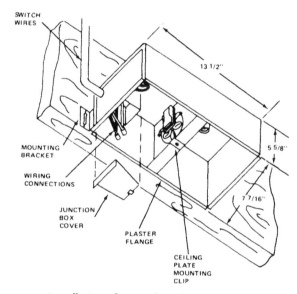

Installation of a simple exhaust fan box.

5. Use a keyhole saw to cut around the box outline.

6. From above the room, install the exhaust fan, attaching it to the adjacent rafter or joist.

7. Follow wiring procedures in Chapter 14 to install the electrical components of the exhaust fan. Some fans require an on-off switch in the circuit while others have automatic switches in them that require no additional wiring.

8. Use a saber saw to cut an opening through the roof or exterior wall for the vent.

9. Install and attach the vent cap.

10. Install flexible ducting between the exhaust fan box and the exterior vent.

Bathroom exhaust systems can be vented vertically through a roof, or horizontally through a soffit (the underside of a roof) or through an exterior wall. The vent cap you install will depend on where it is installed.

Replacing an Exhaust Fan

Replacing an existing exhaust fan unit is relatively simple *if* you buy one of the same size and rating. Because exhaust fans haven't dramatically changed in design over the past 25 years you may find that the new model by the same manufacturer will exactly fit in the old location.

Heads Up! _____

Before drilling and cutting ceiling drywall, make sure you are wearing safety glasses and a mask so the resulting plaster dust doesn't get into your eyes or mouth.

---

3. Place the exhaust fan box over the marker wire and against the closest rafter or joist to determine where it will finally be mounted, then drill holes at the four corners of the box to mark its location.

4. In the bathroom, place the exhaust fan box between the four holes in the ceiling to verify, then mark the perimeter with a pencil. You should have drawn a rectangle with a hole in each corner.

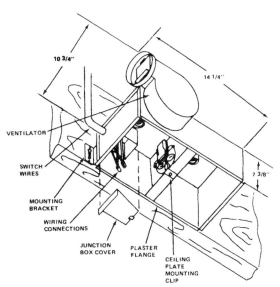

The fan pulls air from the room and out the vent.

It's relatively easy to remove an existing exhaust fan and replace it with a new model.

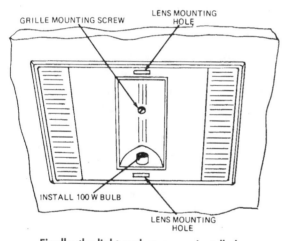

GRILLE MOUNTING SCREW

LENS MOUNTING HOLE

INSTALL 100 W BULB

LENS MOUNTING HOLE

Finally, the light and cover are installed.

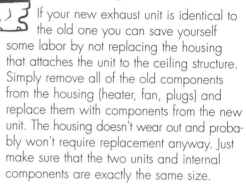

Potty Training

If your new exhaust unit is identical to the old one you can save yourself some labor by not replacing the housing that attaches the unit to the ceiling structure. Simply remove all of the old components from the housing (heater, fan, plugs) and replace them with components from the new unit. The housing doesn't wear out and probably won't require replacement anyway. Just make sure that the two units and internal components are exactly the same size.

As mentioned, the job of installing space heaters, exhaust fans, and other heating and ventilation components in bathrooms is made much simpler if the walls are "open" or the framing is uncovered. Chapter 16 tells you more about walls and flooring systems.

The Least You Need to Know

- Most bathrooms require a supplemental heating and/or ventilation system to keep the room comfortable and the air fresh.
- Space heaters are supplemental heating systems that usually can be installed by the homeowner following the heater manufacturer's instructions.
- Many homes install a combination heating, ventilation, and lighting system in the ceiling for efficiency at lower cost than individual systems.
- Working safely means knowing the inherent dangers of the job you're doing and avoiding them.

In This Chapter

- ◆ Understanding how walls and floors are built and repaired

- ◆ Choosing building materials for your wall or floor job

- ◆ Step-by-step instructions for remodeling a bathroom wall

- ◆ Illustrated directions on building or remodeling a bathroom floor

  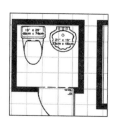  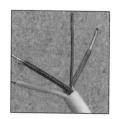 

Chapter 16

Wall and Flooring Systems

If your bathroom remodeling job is more than a facelift, chances are you will have to tear into, replace, or install a new wall. If the floor system is damaged or you're adding a new room to your home as a bathroom you will fix or install a new floor.

This chapter will show you how walls and floors are built and remodeled. In addition, you'll learn how to install a skylight in your bathroom. To make the job easier, I've included numerous hands-on photographs and drawings so you can visualize the job ahead.

Looking Inside Walls and Floors

Ever wonder what's inside the walls in your home? The answer is: mostly air. However, the walls are also home to the plumbing pipes, electrical wires, and insulation.

Before tearing into a wall or trying to build a new one, let's take a look inside the walls of your home. Let's look at the framing, enclosed systems, insulation, and sheathing (covering), as well as the floors below.

Framing

A wall frame is the building component that not only separates one room from the adjacent one, it supports the ceiling and any other rooms above it. A floor frame disperses the weight of the entire house, including all wall frames, and its contents evenly over the foundation. The foundation disperses all weights over the ground on which the house sits.

Walls are made of horizontal plates and vertical studs.

There are two types of walls in a house, each named for what they do: loadbearing or nonloadbearing. A loadbearing wall is one that is important to support the house's weight. If the wall were removed the house above it could shift or even fall in. A nonloadbearing wall is functional, but not as a support. It could be removed without structural damage. Often, single story homes only have the perimeter walls and one long central wall in the center that are designated loadbearing. Two story homes (and homes with heavy roofs) require more support so more walls on the first floor are considered loadbearing.

For support, headers are added above doors and windows.

Potty Training

Determining whether a specific wall is bearing or nonbearing isn't always easy, so you may need a building inspector or contractor to identify them before you try to remodel them. Obviously, it's important to know which is which before you start removing or modifying walls in a bathroom.

Even if you're not installing a new wall you may be remodeling it to include a new window and need to know how walls are framed. Here are the components from bottom to top:

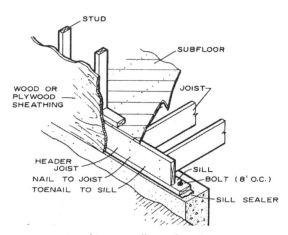
Attaching a wall to a foundation.

- Foundation, typically made of concrete or masonry block.
- Sill plate, horizontal lumber fastened to the foundation as the base for the floor.
- Floor joists, a horizontal frame that sits on the foundation and supports the house.
- Subflooring, a wood sheathing or cover over the floor joists.
- Sole plate, horizontal lumber fastened to the subfloor to serve as the base of the wall.
- Studs, vertical lumber fastened to the sole plate at intervals of 16 in. or 24 in. o.c. (on center, or center to center).

◆ Top plate(s), horizontal lumber fastened to the top of the studs to serve as a cap to the wall.

◆ Corner brace, lumber fastened to studs diagonally between sole and top plates.

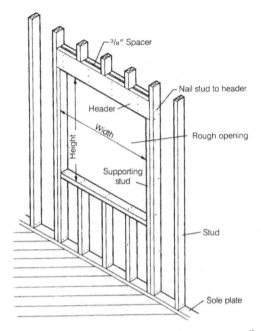

Components of a typical window frame in a wall.

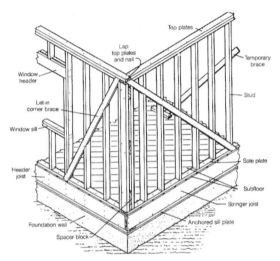

Components of a typical residential floor and wall system.

Because windows and doors require openings in otherwise framed walls, they require special treatment to help disperse the weight of the house. The removed vertical studs are replaced by a larger wood member, called a header, at the top of the opening. Cripple studs (also known as jack studs) assist in supporting the header as well as the window frame. Door frames are built similarly, except there are no cripple studs under the opening.

Because you will be working under and around it you should know how roofs are framed. You may even be adding a dormer to a roof or extending a roof line to add a new bathroom. Many of the components are similar in function and name to those in floor and wall systems.

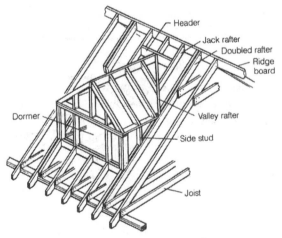

Components of roof and dormer framing.

Enclosed Systems

Once walls are framed and before the insulation goes in and the sheathing goes on, the enclosed systems are installed. These systems include the plumbing, electrical, and HVAC.

Chapter 13 showed you how plumbing systems work and how to install or remodel your home's bathroom plumbing. Chapter 14 covered the same topics for wiring and lighting systems. Chapter 15 explained bathroom heating and ventilation systems as well as showed how they are installed. Each of these systems is enclosed within the wall.

Remember that you will have an official building inspector come to your remodeling site when called to make sure that you are following local building codes. That means you will stop what you are doing and step aside for the inspector to check progress to that point. Depending on your remodeling job you can expect a few inspections. Typical are a frame inspection of all framing, bracing, and utility rough-ins; a wall inspection after the drywall is installed but before it is taped and finished; and a final inspection for everything else.

The point here is to know when these inspections are required and make the appropriate telephone call when you are ready. Don't continue working. If you sheath a wall before the interior has been inspected you'll probably have to tear off the sheathing to allow the inspector a view. In fact, you may have to remove *all* of the sheathing, not just that over the pipes and wiring. So make sure you know what inspections are required and call for them.

Insulation

Once the wall is remodeled and enclosed systems are installed, insulation is added before the wall is covered with sheathing. Insulation slows down the heat transfer process. During summer, it slows down the hot air outside trying to get into the cool house. In winter, it slows down the cold air outside from getting into the warm air inside. The greater the insulation the slower the heat transfer.

There is a wide variety of insulating materials used in home construction and remodeling. The most popular are blanket, batt, fill, reflective, and rigid. Each do about the same job so the selection depends on how easy it is to install in your remodel. Blanket and batt insulation are placed between studs, fill is blown into the wall typically after it is sheathed, reflective insulation is enclosed in walls, and rigid insulation often is installed on the exterior of the wall framing as a sheathing.

> **Bathroom Words**
>
> **R-value** is a measurement of the amount a specific material restricts the transfer of heat. In moderate climates, for example, local building codes require an R-19 rating for exterior walls and R-30 for attic floors. In the coldest climates these numbers are R-22 for exterior walls and R-49 for attic floors. Make sure you buy the appropriate R-value insulation for your bathroom remodel.

Sheathing

Sheathing is simply a covering. For modern walls, that means drywall, a flat gypsum plaster sheet encased in paper. In special applications it can be plaster or stucco, though these are more difficult surfaces for the do-it-yourselfer to install.

Most interior drywall is covered in a gray or white paper. Drywall used in moisture areas such as in bathrooms and showers is sheathed in green (or sometimes blue) paper and is called greenboard. It is relatively waterproof. Later in this chapter, you'll see how to install drywall sheathing including taping.

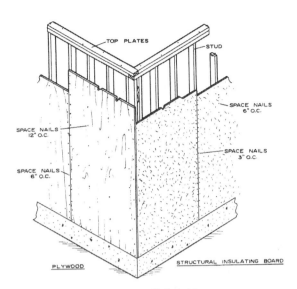

Exterior wall sheathing.

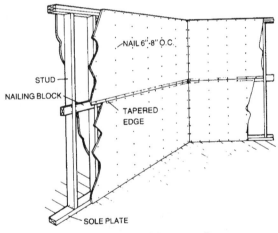

Installing drywall horizontally.

The other task in sheathing is painting the surface. It's a relatively easy task that requires just a few basic tools and materials. In fact, it is one of the most common jobs in bathroom remodeling. Dramatic changes can be made in a bathroom with a facelift that includes painting.

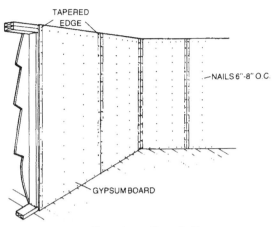

Installing drywall vertically.

Flooring

The floor frame and subfloor are installed or repaired, but they are ugly. Your new bathroom needs new flooring material installed over the subfloor.

The choices in flooring materials are broad. You can select vinyl composition tile (VCT), ceramic tile, stone tile, solid hardwood, or laminate wood flooring. You also can install carpeting, but not near the shower or bathtub.

Which flooring material is best? The greatest dictators of flooring are where you're installing it, who is installing it, and your budget. For floor areas that may become wet—in front of the shower, tub, toilet, and vanity—the best choice is waterproof flooring. If you can install it yourself (most materials) you can opt for better quality for the same price. If your budget is tight and your skills limited you may decide that vinyl tile squares are the easiest and least expensive.

Selecting Materials and Tools

Your remodeling materials list indicates the materials that you will be installing in your bathroom. However, it may not be sufficiently specific. For example, it may indicate 120 sq. ft. of VCT as flooring, but not what grade or design. That's when it's time to go shopping.

So what do you look for in materials? Here are some guidelines.

Lumber

The primary ingredient in framing and remodeling walls and floors is lumber. Lumber is dimensional wood cut to a specified size to make construction and remodeling easier. Most lumber is cut to 8 ft., 10 ft., 12 ft., or 16 ft. in length, with one big exception. Wall studs are 2 in. × 4 in. *nominal*, but cut shorter than 8 ft. They are 7 ft. 8¼ in., just short of the typical height of interior walls.

Bathroom Words

Nominal size is the size designation of a piece of lumber before it is planed or surfaced. If the actual size of a piece of surfaced lumber is 1½ × 3½ in., it is referred to by its nominal size: 2 × 4.

The quality of the lumber you're buying is important, too. The materials list for your remodeling job will probably indicate what lumber grade you should be using. Much depends on where you live and what species and grades are most popular and available in your area. Douglas fir lumber of the west, for example, is graded differently than southern pine or northern fir.

The lumber you buy will be referred to as kiln dried or green. Green lumber is cut and stacked with a high moisture content that can change the shape of the wood as it air dries. Kiln dried wood is put into a heating kiln for a period to remove some of the moisture. Kiln dried is preferred, but is more expensive.

Some remodelers and do-it-yourselfers prefer to use steel wall framing. If local building code allows for its use in construction, ask your building material supplier to demonstrate installation. It's quite easy. In most cases you install the top plate on the ceiling, install the bottom plate directly below it, then install vertical studs in between at 16 or 24 in. o.c. depending on local codes. Steel wall framing is fastened with screws.

Fasteners

Fasteners include nails, screws, bolts, nuts, and adhesives. There are special nails for masonry, roofing, finishing, and other common applications. Nails are classified by the size of the shank and the shape of the head. The most common type is called *common*, with large, flat heads for secure fastening. Next is *finish* nails with smaller heads that aren't so obvious if flush to or below the wood's surface. Nails are sized by length, indicated by a *d* or *penny*. For example, a 4d nail is 1½ in. long; an 8d nail is 2½ in. long.

Bathroom Words

When referring to nails, a **penny** originally indicated the price per pound. It has since been standardized to refer to the nail's length. To determine the nail's length, multiply the pennies by ¼ in., then add ½ in. to the total. For example, a 6d (six penny) nail is (6 × ¼) + ½, or 2 in. long. A 10d nail is (10 × ¼) + ½, or 3 in. long.

Screws are pointed-tip, threaded fasteners installed with a screwdriver. Round- and pan-head (flat-head) screws require a straight-tip screwdriver, and Phillips-head screws require a Phillips screwdriver. Screws are sized by length.

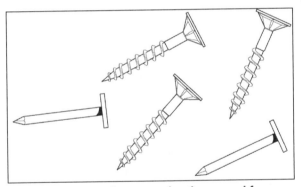

Special drywall screw and nails are used for fastening drywall to studs.

Bolts are flat-tipped, threaded fasteners that use a threaded nut to attach wood or metal together. A washer may be placed under the bolt head or the nut for a firmer fasten. Bolts are classified by the type of head. Stove bolts and machine screws (actually bolts) are turned with a screwdriver. Hexagon- and square-head bolts are held in place with a wrench while the nut is turned to tighten. A carriage bolt's head imbeds itself into the wood when the nut is turned. Bolts are sized by length and thread.

Adhesives secure the surfaces of two materials together. Adhesives come in liquid, solid, or powder form, and some require a catalyst to activate them. Some adhesives are waterproof while others are not; some need to be held together (clamped) while drying and others don't. Select adhesives based on their characteristics, strength, setting time and temperature, and bonding method.

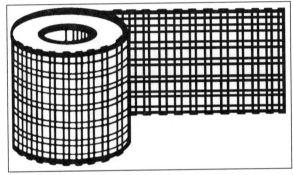

Drywall tape is relatively inexpensive and easy to use.

Tools

You'll need a variety of tools for various wall and floor remodeling jobs. Here's a basic list:

◆ A good quality 8- or 16-ounce curved-claw hammer
◆ A set of screwdrivers or a combination screwdriver with assorted tips (standard and Phillips)
◆ An adjustable wrench (6, 8, or 10 in. long), or a set of wrenches (open- and closed-end) with standard (inches) and metric (millimeter) sizes
◆ Hand or power drill with assorted bits
◆ Measuring tape
◆ Hand, hack, or power saw for cutting wood, plastic, or metal (depending on the blade used)
◆ Paint brushes or paint pads

- Paint roller (frame and cover) and tray
- Plumb line and bob for installing walls
- Floor installation tools depending on the type of flooring material

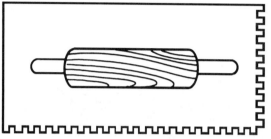

A notched trowel makes spreading a thin coat of flooring adhesive easier.

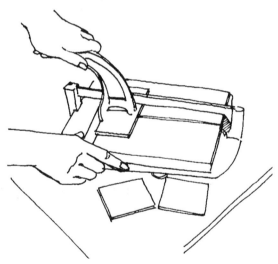

A tile cutter comes in handy for trimming floor and wall tile.

Remodeling Walls

Remodeling a wall means either modifying it, replacing it, or installing a new wall where none existed. For any of these jobs, first review how walls are designed (discussed earlier in the chapter) before proceeding.

Heads Up!

Always make sure that you work safely; remodeling a wall requires tools and materials that can be dangerous. Power tools especially can injure you. Also make sure that all electrical and water systems are completely shut off not only to the room in which you are working but also to adjacent rooms that may share common services.

Framing a Wall

Framing a new wall is relatively easy. Refer to your remodeling plan for location and dimensions as you perform the following steps:

1. Cut the sole or bottom plate to length and fasten it to the subfloor with nails.

2. Use a plumb line to determine the exact location for the top plate directly above the bottom plate, and then fasten it to the ceiling.

3. Cut, install, and fasten wall studs between the top and bottom plates, making sure they are plumb.

4. As needed, install headers and cripple studs for window and door openings.

In new construction where the roof is not yet in place you can build the walls horizontally on the subfloor and tip them up into position. If there is sufficient room and clearance in the new bathroom you can build and tip as well, then install thin wood wedges above the top plate to hold them in place. Just make sure that the walls are square and plumb.

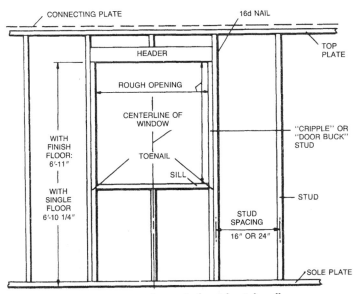

Components of a typical residential wall.

Studs are spaced differently at the junction of perpendicular walls.

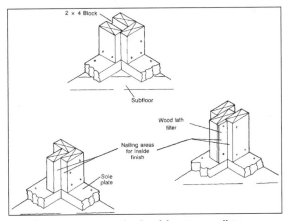

Three common methods of framing wall interconnections.

Installing Insulation

Once the wall framing is in place and the utilities (electrical, plumbing, HVAC) are installed you can install insulation in the open walls.

Flexible insulation simply presses into place between studs and is fastened either with staples or by the pressure of a tight fit. Rigid insulation will stand on its own either between studs or, more commonly, attached to the outside of the studs to serve as both insulation and sheathing.

Batt insulation is installed in framed walls before drywall is installed.

Remember as you install a *vapor barrier* that it should be installed on the *warm* side of the insulation. Otherwise it is less effective.

Bathroom Words

A **vapor barrier** is a material used to retard the flow of vapor or moisture into walls. The two types of vapor barriers are the membrane and liquid. Membrane barriers come in rolls that are attached to the side of the wall. Liquid vapor barriers are applied with a brush.

Installing Drywall

Drywall is a common material extensively used in remodeling bathrooms and other areas of the home. Though it comes in sheets of 4 ft. × 12 ft., most do-it-yourselfers prefer sheets of 4 ft. × 8 ft.

because it is easier to handle. Common thicknesses are ⅜ in. for walls and ½ in. for ceilings, though local building code dictates requirements. Building code also tells you where and how many drywall nails or screws are installed. Most drywallers install ceiling panels before wall panels; ceiling panels typically require more fasteners.

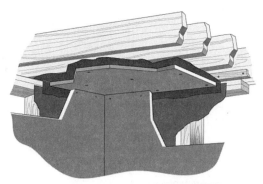

Typically, ceiling drywall is installed prior to wall drywall.

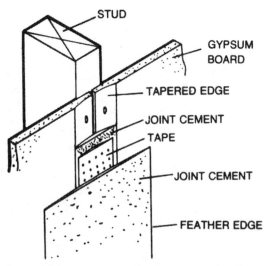

Drywall joints are covered with compound and tape to make them smooth.

Installing drywall is relatively easy. If you need to cut a sheet, first use a straight edge to score and cut the front, then bend the board back and score and cut the paper on the back. Here are basic installation steps:

1. Place the sheet of drywall tightly against the wall frame.
2. Use a pry bar to carefully lift the panel into place against the ceiling panel.
3. Nail or screw the drywall sheet into place attaching it to the wall studs at specified intervals.

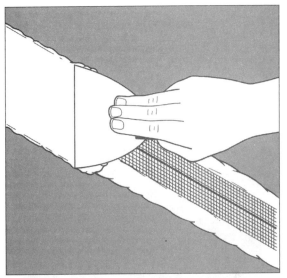

Apply a final coat of compound and let it dry before sanding smooth.

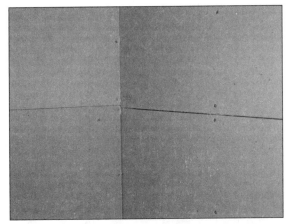

Drywall typically is installed horizontally with tapered edges butted.

4. Apply drywall joint compound ("mud") in the channel at the joint between two sheets of drywall.
5. Apply drywall tape over the joint compound and smooth it with a drywall knife.

Make sure you firmly press the drywall tape into corners.

Place the drywall tape over the drywall compound.

6. Apply a final coat of drywall joint compound over the drywall tape and smooth it.

   Let the joint dry, then carefully sand it flat before texturing and/or painting the drywall.

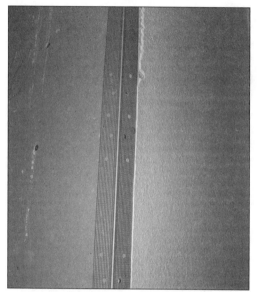

Drywall corners must stand up to abuse, so install metal edging.

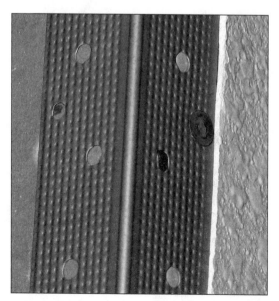

Drywall corner edging is installed by nailing it through the drywall and into the supporting studs.

Installing Doors, Windows, and Skylights

Doors, windows, skylights, and other barriers are installed after the interior walls are sheathed or closed. Fortunately, modern doors and windows are prehung, purchased as a package that includes casing and trim, making installation easier. Here's how to install a prehung door:

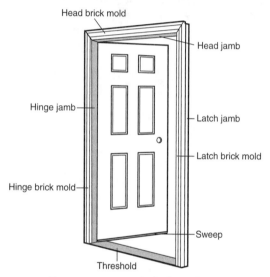

Door components including trim.

1. Verify that the rough opening is correct (instructions with the door include specific dimensions).

2. Make sure that the opening is square and level.

3. Unpackage the door assembly and remove the skid (bottom) board used for support in transit.

4. Based on the manufacturer's instructions, prepare the door's threshold now or once the door is installed. Also remove any trim as indicated by the instructions.

5. Center the assembly in the opening and tilt it up into position.

6. Plumb the door frame on the hinge side, using screws to temporarily fasten the frame to the stud.

7. Plumb the door frame on the latch side, using screws to temporarily fasten the frame to the stud.

8. Insert wooden wedges or shims behind the hinges and other points around the frame for support.

9. Make sure the door frame is plumb and square, and then finish attaching the frame to the studs following the manufacturer's instructions.

10. Remove any shipping clips that hold the door closed.

11. Install and/or adjust door locks and hardware.

12. Install insulation as needed between the opening and the door assembly.

13. Install trim.

14. Apply caulk around the door.

15. If not previously required, install the threshold.

Installing prehung windows and skylights is similar to that for doors. Once the opening size and squareness are verified, install the prehung units, fasten them to the frame, remove shipping clips, and add the trim. When installing a skylight, think of it as a roof window rather than a wall window.

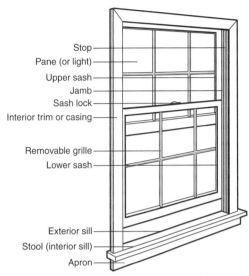

Window components including trim.

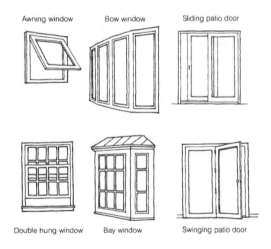

Popular window arrangements.

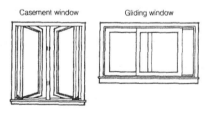

Potty Training

Make sure you read and follow the manufacturer's instructions when installing doors, windows, skylights, and solar tubes. They will have more specific instructions about sealing and protecting them.

Prehung windows are slipped into frame openings
in walls, then attached.

Once the window is installed, trim will cover the
framing.

Skylights are horizontal (or nearly so) windows. The
frame is built and the prehung skylight installed.

Remodeling Floors

Remodeling a bathroom floor can be as simple
as laying new flooring over old. Or it can require
replacing subflooring or even floor joists. Structural repairs shouldn't be done by the inexperi-
enced, so let's stick with subfloor repair and
laying new flooring.

Subfloor repair typically means removing
the damaged (often by moisture) subfloor and
replacing it with new material. As you cut out
the old flooring, make sure the circular saw
blade used isn't set too deeply that it cuts more
than ¼ in. into the floor joist. Also make sure
that any remaining and new flooring end at the
middle of a joist for support.

Most tile requires a firm surface, such as backerboard, below it.

You can install leveling products over existing flooring
materials.

Leveling products require smoothing with trowels to get a level surface on which tiles will be installed.

How you install new flooring material depends on what the material is. One thing all flooring has in common is that it needs a clean, level surface. Thinner flooring such as VCT will transmit any imperfections in the subfloor to its own top surface. Thicker flooring such as hardwood can absorb minor variations, but not uneven subfloors.

Tile adhesive is spread over the dried leveling product.

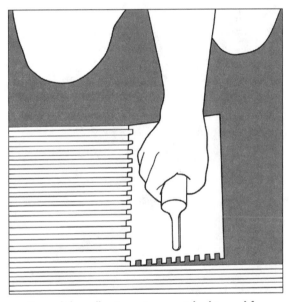

Spread the adhesive using a notched trowel for greater adhesion.

Ceramic tile is a popular flooring material for bathrooms, but it can be cold to the touch.

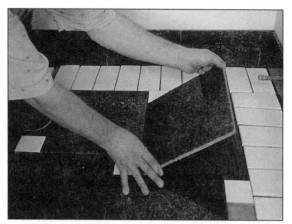

Set the tile in place at one edge and then lower it into position on the adhesive.

Spread grout, working it into the space between the tiles.

The best advice for installing flooring material is to follow the manufacturer's directions. Some flooring manufacturers offer installation kits for products they sell directly to consumers. Also, buying your flooring materials from a flooring store or department often can get you assistance from an experience clerk. Ask lots of questions.

Once the whole tiles are installed, measure and cut the edge tiles.

The Least You Need to Know

- Framing is the skeleton of the wall, ceiling, and floor.
- Framing materials are standardized to make the job easier—even for the do-it-yourselfer.

- Before tearing out a wall, make sure you know whether it is loadbearing or non-loadbearing.
- Make sure you know the stages at which building inspectors require you to stop so they can check progress in wall construction.

In This Chapter

- ◆ Taking a look at how cabinets are built
- ◆ Choosing the right cabinets for your remodeled bathroom
- ◆ Installing cabinets and countertops
- ◆ Refinishing cabinets for low-cost remodeling

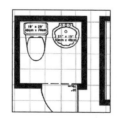

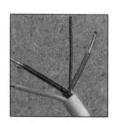

Installing Cabinets

Cabinets are a vital component of any bathroom remodel job. Even a bathroom facelift can be enhanced with a new set of cabinets. Fortunately, premade cabinets are readily available and easy to put in. Countertops, too, are easy to install even for the first-time do-it-yourselfer.

This chapter shows you how to select and install bathroom cabinets from the floor up. It illustrates the differences between various materials and guides you in selecting the right cabinets for your remodel and your budget.

Selecting Bathroom Cabinets

How are bathroom cabinets made, anyway? There are all those doors and drawers, sliders and glides. How do you choose the right units for your bathroom remodel?

First, know that though cabinetry is an art, selecting premade cabinets is a relatively easy task. It is one of the most common remodeling jobs that do-it-yourselfers tackle and so manufacturers have made it easy. Even professional cabinet builders often install premade cabinets because they fit their client's budget.

To get a better sense of how to install bathroom cabinets, let's look at typical configurations, sizes, materials, and countertops.

Cabinet Construction

Cabinets are empty boxes to be filled with sinks, drawers, and things to store. The component that holds it all together is called the frame. The bottom is called the base floor. The front is called the face, made up of horizontal rails (top, intermediate, bottom) and vertical stiles. Doors are hinged from the stiles and drawers slide in and out on runners or glides. Some cabinets have backs while others are open-backed. All have a countertop into which the sink and faucet are installed. The lowest part of the base is recessed, called the toekick, to allow you to get closer to the front edge of the cabinet.

Cabinet systems are made up of separate bases and tops assembled in the new bathroom.

The two common types of hinges are framed and frameless. Framed hinges have a hinge bolt exposed while frameless (called European) hinges don't. Framed hinges come in various shapes depending on whether the door is mounted into or in front of the cabinet frame. Cabinet latches include friction, roller, magnetic, and touch latch.

Configurations

The most common configuration for bathroom cabinets is a base cabinet sufficiently wide for a single sink and some countertop room. The minimum width typically is 30 in. To have two sinks, cabinets need to be 60 or more in. wide.

Besides width, the lavatory configuration you choose will depend on how much and what type of storage you need. For example, how many drawers do you require? Four is the standard number in a base drawer cabinet. How about under-sink storage shelves? Remember that the sink and plumbing will take up some of the storage space. Do you need rollout shelves or are stationary shelves okay?

When planning cabinetry configurations, remember to allow for clearances. For example, make sure that drawers can be fully extended without hitting other objects. This may take special planning if you install adjacent cabinets at a 90 degree angle to each other.

Cabinet Sizes

Bathroom cabinets seemingly come in a wide variety of designs. However, whether you're remodeling your bathroom with stock (pre-built and purchased through a store) or custom (built specifically for your bathroom) you'll find that the dimensions are standardized to make construction and installation easier. Here are some basics:

◆ Base cabinets are those that are attached to the floor and wall.

Potty Training

Actually, cabinets are available in *any* size you want them—for a price. Standard cabinets will fit most installations, but custom cabinets are available through local cabinet shops. Custom cabinets will not only fit the exact space but they won't look like they came from a building supply store.

- Sink base cabinets typically are 29 in. high and 21 in. deep.

- Single-unit (door, drawers, or both) base cabinets are sold in widths of 12, 15, 18, and 21 in.

- Double-unit (vanity) cabinets are sold in widths of 20, 33, 36, 42, 48, and 60 in.

- Various single- and double-unit base cabinets can be attached to make a single vanity cabinet.

- Wall cabinets typically are 24 in. high and 12 in. deep.

- Wall cabinets are sold in widths of 18, 24, 33, and 36 in.

- Specialized cabinets (corner, island, full-length) are variations to base and wall cabinets that share many of the standard dimensions.

In addition, base cabinets may use fillers, narrow boards installed between the faces of two adjoining cabinets to fill the space and extend the cabinet width. Fillers also are available for corner cabinets.

Cabinet Materials

Bathroom cabinetry, whether custom-made or factory-made, is constructed following proven construction techniques. In addition, they are made of common materials that can withstand the moisture inherent in bathrooms. Here are some guidelines for selecting cabinets with popular materials:

- Hardwood cabinets typically are made with hardwood frames and veneered plywood surfaces, costing more than other cabinet materials.

- Hardboard cabinets use *hardboard* material for doors, backs, and sides; the frame typically is of softer woods such as southern pine.

Bathroom Words

Hardboard (also known as fiberboard) is a manufactured sheet of wood fibers joined together under pressure. **Particleboard** (also known as chipboard) is a composition board of wood particles bonded with a synthetic resin.

- Particleboard cabinets use *particleboard* material for all components, though some use a soft wood frame. The particleboard is finished with a wood or plastic veneer to simulate solid wood.

New cabinet materials will show up over the next few years, but most are simply variations to existing materials intended to save manufacturing time while giving you value.

Countertops

Bathroom cabinet countertops can be purchased premade or can be built on-site once the cabinets are installed. Countertop material includes solid, preform, laminate, ceramic tile, marble, granite, and stainless steel. Let's take a look at each:

- Solid countertops are manufactured sheets of acrylic or polyester resin and color additives.

- Preform countertops are manufactured as a single unit applying a laminate sheet over a particleboard form that includes both the top and the *splashboard*.

Bathroom Words

A **splashboard** (also called the backsplash) is the vertical board at the back edge of a countertop designed to minimize splashing liquids onto the adjacent wall.

- Laminate countertops are manufactured as particleboard with a laminate sheet, designed for final assembly of the splashboard and edging on site.

- Ceramic tile countertops apply tiles over a solid wood base, fill the space between tiles with grout, and seal the grout against moisture.

- Sheets of marble and granite can be installed as solid countertops, polished in the factory or on site.

- Stainless steel countertops are manufactured to size and installed over the cabinet frame on site.

Installing Bathroom Cabinets

As mentioned, bathroom cabinets are relatively easy to install. Just make sure that the floor is fairly level, the walls are plumb, and that you've carefully measured and planned out installation. Also, don't be afraid to do a *dry* installation; that is, set all the cabinets in place to make sure everything fits correctly before installing the first cabinet. Doing so can save you time—and maybe money. In any case, it will give you confidence that your installation efforts won't have to be undone.

Remember to read through the following instructions, as well as any instructions that come with the cabinets you purchase, before starting the job. Also make notes in your bathroom remodel notebook to help you make the installation go easier.

Gathering Tools

What tools will you need for installing bathroom cabinets and countertops? Following is a starting list. Depending on the type of top you

are installing you may require specialized tools for setting ceramic tile or installing sheet laminate.

- Hammer
- Drill
- Screwdrivers
- Level
- Clamps
- Stud finder
- Rotary saw
- Jigsaw
- Caulk gun and silicone caulk
- Finish nails
- Screws
- Safety equipment

Heads Up!

Remember to work safely. If you are using saws remember to wear eye protection and, if you are sensitive to dust wear a face mask. Take your time.

Installing Base Cabinets

Here's how to install the typical bathroom base cabinet:

1. Find and mark the location of wall studs that will be behind the base cabinet.

2. Mark a level horizontal line across the wall at the height of the cabinet.

3. Set the cabinet in place and install shims under the base to level it against the horizontal back line and a level on the front edge.

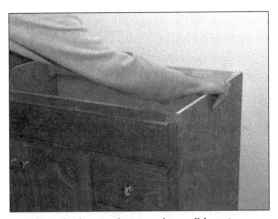

Place the base cabinet at the wall location.

Adjust the cabinet placement to be level and firm.

4. Drill pilot holes and install screws to attach the back of the cabinet to the marked wall studs.

Attach the cabinet to wall studs.

5. Chisel away excess shims that may show.

6. If there is an additional cabinet, abut it to the first cabinet and clamp the faces together, then drill pilot holes and install screws to attach the two cabinets. Level the cabinet with shims and attach the back to the marked wall studs.

It's really that easy. Depending on the type of cabinets you purchased there may be additional trim to install, such as an end panel. Follow the manufacturer's directions for doing so; typically this means placing the panel and either gluing or fastening it. If glued you'll need to apply clamps so it dries flat.

Installing Countertops

The installation of countertops can vary because there are many types of tops. Solid tops such as preform counters require less work than ceramic and laminate tops. In fact, depending on the cabinet you purchase you may be able to buy a countertop that requires nothing more than attachment to the base.

What all countertops have in common is that they need to be prepared for the sink. Solid countertops require that a hole be cut through the top. Built-in-place countertops have a hole cut in the underlayment, and then the surface (ceramic tile) is installed. Here are the steps to installing a typical countertop:

1. Measure the base cabinet length and add 1 in. for the overhang.

2. Cut the countertop to length. If the surface is finished, cut from the bottom side to minimize chipping of the surface.

3. Place the sink in position and draw a cutline inside of the perimeter. Alternately, some sink manufacturers include a template for cutting the sink hole.

Potty Training

If you're working with a large, heavy countertop make sure you protect the corners from damage just in case you bump or drop it. Tape some cardboard protection at each vulnerable corner.

Place the countertop on the cabinet.

4. Drill a pilot hole in a corner of the outline, then use a jigsaw to cut the perimeter line.

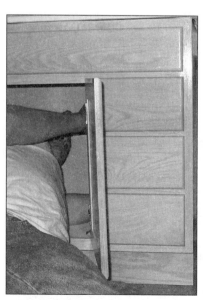

You may need to crawl under the cabinet to attach the top.

5. Install laminate surface and edging as needed, depending on the type of countertop you're installing. If ceramic, install the underlayment before installing the tile.

6. If necessary, install the splashboard.

Tile splashboard is installed with spacers that are later replaced with mortar.

7. Align the countertop on the base cabinet(s) and fasten it from below using corner plates and screws.

8. Install the sink in the hole, caulk it, and attach it from underneath.

9. Caulk seams as needed for a tight, waterproof perimeter seal.

In some cases you might need to attach two countertops, such as a long top or an L-shaped top. To do so, apply caulk between the two pieces and use joint-fastening bolts (available at larger hardware stores) to tighten the joint.

If you are installing a ceramic tile countertop, the underlayment of particleboard or plywood is the base for the tile and requires full installation to the cabinet. You can install an underlayment as support for the splashboard or you can install the tile directly on to the wall. Here are the steps to install a tile countertop:

Chapter 17: Installing Cabinets   **195**

3. Apply adhesive to the underlayment following manufacturer's directions.

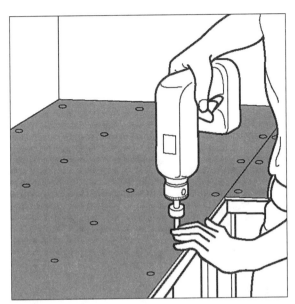

Install a firm underlayment before installing counter-top tile.

Install tile over the adhesive, making sure you place spacers between the tile.

4. Place the first tile and insert tile spacers between tiles as you continue placement.

5. At the rear edge, use a tile cutter to trim tiles as needed to fit the available space.

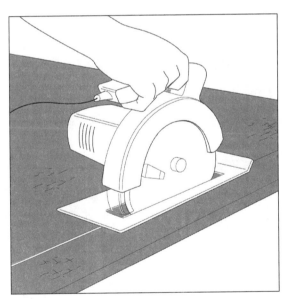

Trim the underlayment to the edge of the counter, being careful not to cut into the counter itself.

1. Determine the location of the first or front row of tiles and mark the back edge.

2. Use a straightedge to draw lines on two sides of the first tile. These marks will align all subsequent tiles.

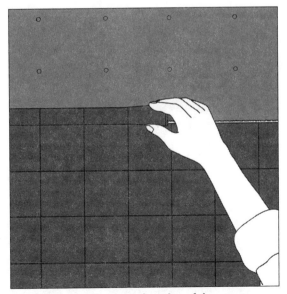

Install trim tile along the edge of the counter.

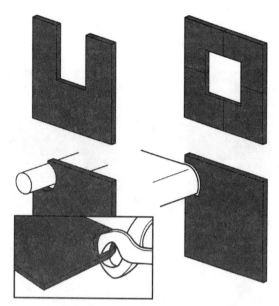

Special trimmers are available from rental centers for cutting tile.

6. Set the tiles by carefully pounding down on tiles using a mallet and a wooden block wrapped in cloth.

7. Remove the tile spacers.

8. Spread the grout mixture (following manufacturer's directions) over the tiles and into the open space between them.

Apply grout, working it between the tiles.

9. Wipe away excess grout and let the grout dry following manufacturer's directions.

Use a squeegee to force grout into the spaces between tile.

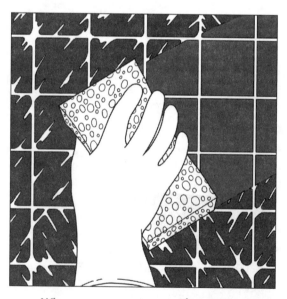

Wipe away excess grout with a sponge.

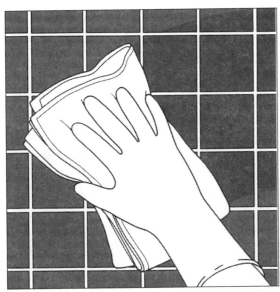

Once done, clean the tile.

10. Once the grout is fully dry, apply penetrating sealer to the grout.

11. Caulk seams as needed for a tight, waterproof perimeter seal.

Caulk the backsplash to seal water from penetrating.

Once the cabinet and countertop are installed you can install fixtures.

Installing Wall Cabinets

Installing bathroom wall cabinets are even easier than installing base cabinets. Typically, wall cabinets are single units that simply attach to the wall studs and ceiling rafters.

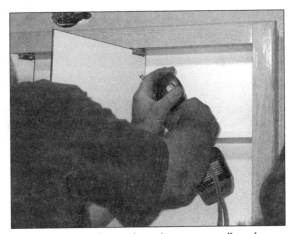

Once level, fasten the cabinet to a wall stud.

If you are installing larger or heavier cabinets, consider installing them *before* you install the base cabinets. You will find it much easier to lift them and hold them in place if you're not working over existing cabinetry. Smaller cabinets can be installed relatively easily over existing cabinets. In fact, you can support the wall cabinets with a temporary frame built to sit atop the counter. If you are concerned about damaging the counter surface, either protect it or use the frames prior to installing the final counter surface.

Waste Lines

Don't be afraid to ask for extra help, especially when installing heavy cabinets. An extra set of hands—and muscles—can not only make the job go faster, but safer.

Refinishing Existing Cabinets

You have another good option: simply refinishing the existing cabinets. If you're tired of the cabinets but the frames are sound, refinishing can be the best idea. It's also cheaper than buying new cabinets so it fits nearly everyone's budget.

Refinishing wood cabinets requires sanding and preparing then applying finish. However, many modern bathroom cabinets aren't solid wood, they are laminate or even faux-wood. How can you tell what your cabinets are made of? Find a finished surface that doesn't really show and scratch it with a sharp object. If the surface is merely a photograph of wood you'll scratch right through it. You won't be able to sand it down and refinish it. However, you can cover it with paint or other resurfacing products.

Once your bathroom cabinets and countertops are installed you can start planning the installation of fixtures—faucets and water controls. That's the topic of the next chapter.

The Least You Need to Know

- Bathroom cabinets are constructed for ease of installation as well as functionality.

- Make sure you dry fit all cabinets to verify that they fit exactly as planned before installing them.

- Countertops are often more difficult to install than cabinets, but can be made easier by purchasing preformed tops.

- If the frames are sound, consider refinishing your existing cabinets to save money.

In This Chapter

◆ A look at how lavatory, bathtub, and shower faucets work

◆ Installing bathroom lavatory faucets

◆ Installing bathtub and shower faucets

◆ Installing bathroom drain system components

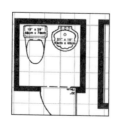

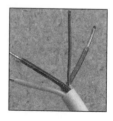

Chapter 18

Installing Fixtures

Once the plumbing is roughed in, wiring completed, heating and ventilation systems in, walls and floor finished, and cabinets installed it's time to finish the plumbing. That means installing the final components of the plumbing system.

Chapter 13 showed you how to install plumbing pipes, Chapter 10 offered simple instructions for installing a toilet, Chapter 11 covered shower and tub installation, and Chapter 12 included coaching on installing whirlpool tubs and specialty showers. What they *didn't* cover are the faucets and water controls. Because there are so many types they get their own chapter—this one.

Understanding How Faucets Work

Plumbing fixtures include tubs, showers, and sinks *and* their water controls. During many remodels installing tubs, showers, and sinks is considered part of rough plumbing and adding the faucets and other water controls is the last step before you get to turn the water on.

A *faucet* is a terminal valved outlet. That is, it is the last component in a water system and it is built around a *valve*, a device that regulates the flow of water.

As you know, there is a wide variety of faucets. Lavatory faucets serve the sink, bathtub faucets control water entering the tub, a shower head and controls are used to mix and deliver water to the shower. There are several types of valves—compression (sometimes called stem-and-seat), disk, ball, and cartridge are the most common. Faucets also use different configurations of levers and handles to open and close the valves. That's why they look different. Some faucets have a spout separate from the valves, such as in many tubs, while others combine both hot and cold valves and the spout into a single unit, such as on many sinks.

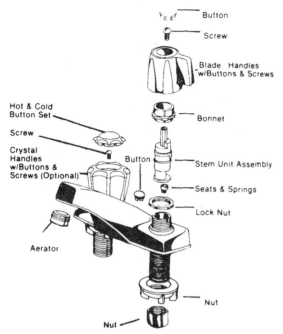

Components of a typical compression faucet.

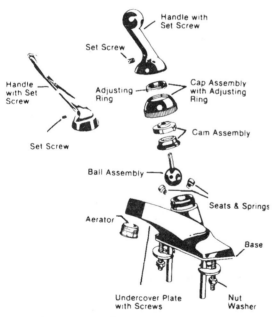

Components of a typical cartridge faucet.

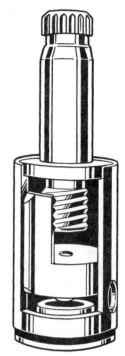

Cross-section of a faucet cartridge.

A related fixture is the drain system. Most are relatively simple. They are located in the lowest part of the sink, tub, or shower and designed to allow either a tight seal against water flow or a quick flow of water when you're ready to drain it. Most drains include a lever system that allows you to control the seal.

Lavatory Faucets

The components of a typical lavatory faucet include the shutoff valves located on the wall and below the sink, the water supply tubes, the hot and cold water valve(s), and the spout. Related components include the drain and drain pipe.

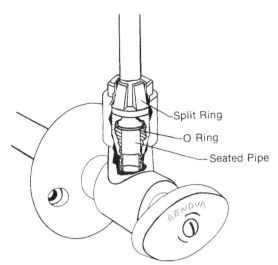

Components of a typical shutoff valve.

A compression valve faucet has two handles that *rise* when you turn the water on. If the handle *lowers* when water is turned on it's a reverse-compression valve. The hot and cold water valves may be attached to the spout on each side of the faucet or they may be independently mounted on the sink. If the faucet only has one handle in the middle it is a cartridge valve and faucet. The hot and cold water are mixed inside the cartridge based on the position of the control lever.

Potty Training

Remember, you usually get what you pay for. Don't buy cheap faucets and fixtures and expect them to look as good or last as long as better quality products. Open the package before buying and inspect the materials and workmanship. Especially beware of fixtures that are shrink-wrapped so you can't easily open them for inspection.

Most lavatories include a stopper to control water flow from the sink into the drain system. The components include the opening or flange, stopper, and lift rod. Lifting the lift rod (located in or behind the spout) activates mechanism in the drain pipe to pull the stopper down and seal the flange. Water is held in the sink until the lift rod is lowered. It then flows through the trap and drain.

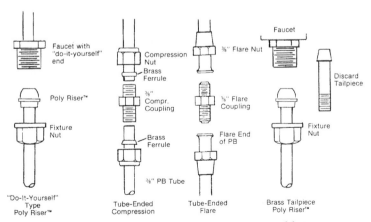

Water supply lines interconnect using a variety of fittings.

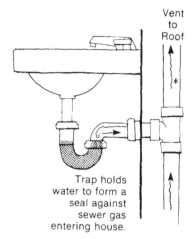

Trap holds water to form a seal against sewer gas entering house.

How a sink trap works.

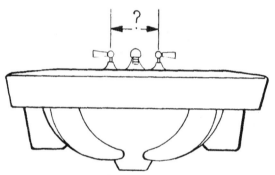

Make sure that the faucet you purchase matches the holes in the sink.

Bathtub Faucets

Bathtub faucets operate similar to other bathroom faucets except that they typically are larger for more water flow. In addition, most are installed with a separate spout and control system. The controls may be combined in a single cartridge or mixer valve or they may be separate valves on either side of the spout.

Bathtub drain systems operate similar to those in lavatories. A lift rod, usually located on the spout, operates the stopper against the drain flange.

Shower Faucets

Visit a large building or plumbing supply store and you'll see dozens of shower faucets and controls in all shapes and sizes. The choices can be confusing. In addition to hot and cold water valves and mixers there are numerous ways to control the water that comes from the shower spout, called the *head*. The shower head can be stationary or hand-held, steady or pulsating, single or multiple. However, most work on the same principles to control the water flow. They are diffusers.

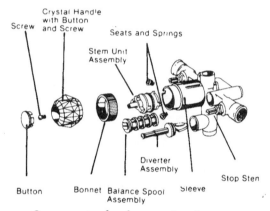

Components of a shower mixer faucet.

Pressure drops are common in home plumbing systems. Someone turns on a faucet elsewhere in the house or yard and all of a sudden someone in the shower gets *very* hot water. The solution for this problem is a pressure-balance spool. It is a special faucet valve that regulates pressure drops to balance output. No more jumping around in the shower. Installing pressure-balance faucets is similar to the task for most other faucets. They can be installed at any faucet in your home, but make most sense in the shower.

Shower drains are relatively straightforward. Because you don't want to keep water in the shower base, the drain remains open at all times.

Typical shower drain. Make sure you cover it while installing faucets as you could lose small parts down the drain.

Installing Faucets and Controls

Chapter 13 showed you how bathroom plumbing systems work and how rough plumbing is installed. Rough plumbing, remember, is all the pipes that go into the walls and under the floor. Once it is completed you install wiring, walls, cabinets, and sinks, then faucets. Rough plumbing often includes installing bathtubs and showers though not the faucets. That's what you're doing now.

In Chapter 13 you installed various types of pipe in the walls for the lavatories, bathtub, shower, and toilet. You'll now install the faucets and other water controls for the lavatories, tub, and shower.

Faucets are the termination of your plumbing supply system. Drains are the beginning of the DWV (drain-waste-vent) system. Once the sink, tub, and shower are installed it's relatively easy to install these components with basic tools.

The tools you need for installing faucets and drains include basic hand tools: wrenches and screwdrivers. Depending on what else you're simultaneously doing you may need a pipe wrench and other specialized tools such as Allen wrenches. Fortunately, many faucet kits include the special tools you need.

Materials needed are those on your material list. However, you often can change your mind. You may decide that the gold-plated faucet is a little ostentatious (and expensive) and opt for bronze. Or you can replace one type of faucet with another—as long as the fixture mounting holes are the same. You still have choices.

Installing Lavatory Faucets and Drains

Fortunately, most lavatory or sink faucets are sold in packaged kits with all the components, instructions, and even specialized tools included. To make the job easier, some include instructions on the back of the package to show you how they are installed—good information to have *before* you buy.

Here are the typical steps to installing lavatory faucets:

1. Install the shutoff valves on the hot and cold pipes.

The shutoff is a small valve that stops water flow if you need to replace a line.

2. Place the faucet on the sink or counter to verify alignment in the holes. If there are no holes, measure and cut them into the countertop before proceeding.

Make sure the faucet fits the openings in the sink before installation.

If the faucet has plastic nipples, make sure you use plastic fittings.

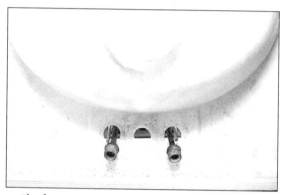

The faucet connections will protrude through the openings in the sink.

3. Install the faucet in the mounting holes using provided gaskets or plumber's putty.
4. Attach the faucet(s) to the countertop or sink with provided locknuts.
5. Install supply lines from the shutoff valves to the underside of the faucet, tightening them against leaks.

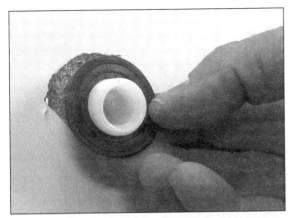
For a better seal, make sure you install proved washers.

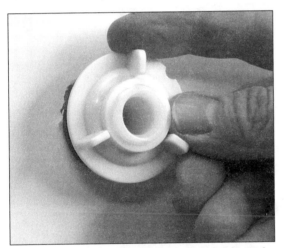
Plastic locknuts are intended to be hand-tightened.

Ends of a water supply line.

The process is a little more complicated if there is a separate hot water valve, cold water valve, and spout, but these general directions will guide you. The job is still relatively easy, especially if you plan to take your time.

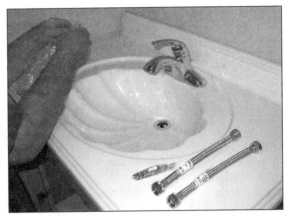
Lay out all parts to make sure you have what you need before installation.

Installing the drain and trap requires a little more skill, but is well within the range of most do-it-yourselfers. What you're doing is connecting up the sink flange with the drain pipe that exits the wall. The only tricky part is installing the trap. Here's the process:

1. Insert a short pipe, called the tailpiece, into the drain spud.

The tailpiece (right) is connected to the drain line (center) by installing a trap.

Install the tailpiece.

2. Align the trap-to-drain line with the drain.

The tailpiece extends from the sink spud.

Install the trap.

3. Mark the bottom of the tailpiece where it needs to be trimmed to fit inside the trap.

4. Cut the tailpiece to the correct length.

5. Install and tighten the tailpiece, trap, and drain line.

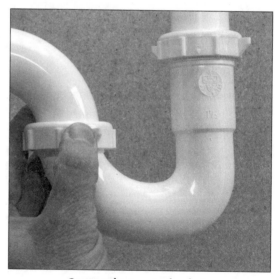

Connect the trap to the drain.

Other joints can be installed in the same manner to serve a second sink.

Most PVC pipe is intended to be hand-tightened.

If you are installing a stopper you'll need a special tailpiece that includes a lever for moving the stopper inside the pipe. Otherwise, water would leak out of the hole in the tailpiece. Stopper systems include installation and adjustment instructions.

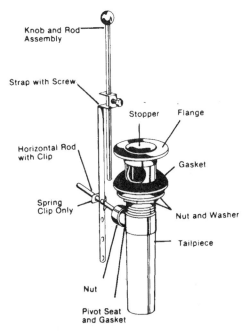

Components of a typical sink stopper system.

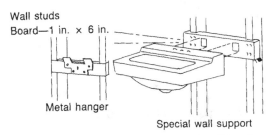

Some lavatories are hung on special wall brackets instead of mounted on cabinets.

Heads Up!

When installing sinks and fixtures, use a washcloth or masking tape to cover drains until all components are installed. Invariably, a small, hard-to-replace part will fall down the hole and be difficult to retrieve. Better safe than sorry!

Installing Bathtub Faucets and Drains

Bathtub faucets, too, typically come with full instructions for installation. In fact, you shouldn't buy one without clear instructions. Here's a typical installation:

1. Install the valves on the hot and cold water supply lines. Some bathtub faucets have a single mixer valve instead.

2. Install the tub spout, typically screwed on to the end of the outlet pipe. Some have an Allen set nut on the underside to cinch the spout so it doesn't continue to rotate.

If the bathtub is a tub and shower combination, refer to the following directions on installing the shower head.

Installing the stopper in a bathtub drain system can be a little tricky, depending on access. If you've allowed for access from behind the end of the tub the job is easier. Easiest of all is to install the stopper and control mechanism *before* installing the tub, as suggested in Chapter 11.

Installing Shower Controls

Shower controls include the valve(s) that manage the mixture of hot and cold water to the shower head. If the shower is part of a bathtub the controls typically will be the same for both with a diverter valve that selects whether water goes to the shower head or the bathtub spout. Most tub/shower diverters are built in to the bathtub spout so there is no additional installation.

Shower valves are installed just as those in a bathtub. The hot and cold valves or a single mixer valve is installed on the end of the appropriate pipe.

Installing Shower Heads

Shower heads are simply faucet spouts. Some have additional controls to regulate the spray. In any case, they are very easy to install. The process is simply to thread the head on to the supply pipe extending from the wall (added during rough-in, Chapter 13).

Typical shower head and controls.

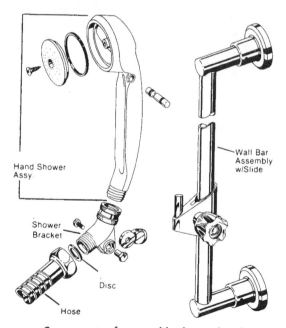

Components of a movable shower head.

Remember to slip the flange cover over the pipe before installing the head.

Applying plumber's tape for a tighter seal.

The Least You Need to Know

◆ Bathroom faucets are valves designed to control the flow of hot or cold water.

◆ Faucets and drains are relatively easy to install and most come with clear instructions.

◆ Lavatory and bathtub drains require planning for trouble-free installation.

◆ Make sure you carefully install and adjust stoppers following manufacturer's directions.

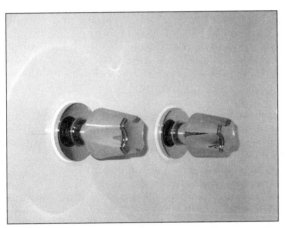

Most shower controls have a set screw that securely fastens them to the valve stem.

Remember, however, that many shower heads are made of relatively brittle materials. Wrap the end of the pipe with plumber's tape for a tight seal and don't overtighten the head when you install it.

In This Chapter

- ◆ Painting walls and other surfaces
- ◆ Adding wallpaper to your remodeled bathroom
- ◆ Installing trim
- ◆ Installing accessories

  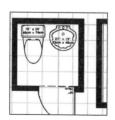  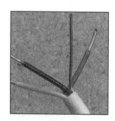 

Chapter 19

Final Touches

Whether you are thoroughly remodeling your bathroom—removing walls and fixtures—or simply giving an old bathroom a facelift, this is a very important chapter. In it are illustrated instructions for dramatically enhancing the look of your home's bathroom with paint, wallpaper, trim, and accessories.

Early chapters gave you design ideas for your remodeled bathroom. This one helps you convert those ideas into reality. Even so, feel free to change your mind. If you were going to paint but now think you'd rather wallpaper, go ahead. Or if you want to add more decorative accessories, have at it. You won't need an addendum to your building permit. All you need to do it is *do it*.

Painting Your Bathroom

Painting a bathroom is about as easy a job as it gets. Modern paints and tools make not only application easier, but cleanup is relatively painless. Just follow a few simple rules, outlined here, and enjoy the dramatic difference in your remodeled bathroom.

One safety tip: Make sure you have adequate ventilation in the room you're painting. Even latex paints give off fumes that can harm you if concentrated in a small room. Keep the vent fan on or a window open. If necessary, install a fan at the door that disperses the fumes.

Selecting Paint

Paint is more than color. It's a protective wall coating, too. So selecting bathroom paint first means selecting the type of paint you will use, based on its properties. Oil-based (alkyd resin) paints seem richer, but are moderately more difficult to apply and clean up. Latex paints are easier to apply and clean, and are available in various glosses.

Interior paints are oil-based, latex, and varnish paints. The major difference is the primary ingredient, called the carrier. Oil-based paint uses linseed or a similar oil as its carrier. Latex paints are water-based. Varnishes are solvent-based; they use mineral spirits, alcohol, or other solvents to carry the color.

The paint you select also depends on what it is going to cover. Masonry paint has very different properties from paint used to cover drywall or wood. Application is slightly different, too. Fortunately, paint containers offer information on what they cover and how to apply the paint.

Potty Training

As you select paint, make sure you look for enamel and other coating types that resist penetration by moisture. Some paints specifically say they are formulated for bathroom surfaces.

How much paint will you need? Depending on whether you're first applying a primer coat and how porous the surface is, most paints suggest that a gallon can cover about 400 sq. ft. That's approximately four walls in a 10 x 15 ft. room. Actual coverage can range from about 300 to 500 sq. ft. per gallon. The paint container will offer guidelines and an experienced paint store clerk can give you advice.

What colors should you choose? Many bathrooms don't have much natural light. Even artificial light impacts how colors are seen. Here are some tips for selecting paint colors based on natural and artificial light:

- Smaller rooms seem larger if all surfaces are of the same color.
- Accent a room by painting one wall a color compatible with that on the other walls.
- De-emphasize a long room by painting the walls on opposite ends with a darker color or shade.
- Apply cool colors (green, blue, violet, gray) on walls that receive warm natural light such as from the south or west.
- Select warm colors (red, orange, brown, yellow) for walls that get cool natural light from the north and east.
- Buy samples of selected colors and apply them to a section of a wall to see how they look in the room with installed lighting.
- Select colors based on the use and desired mood of the room; your paint store can help.

Selecting Tools

You don't *need* many tools to paint a bathroom. A brush is about all that's necessary. However, you can make the job easier with a few other inexpensive tools:

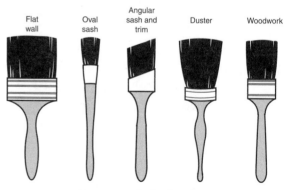

Common painting brushes.

- A paint roller and tray will let you put more paint on the wall faster.

- A roll of plastic drop-cloth can make cleanup easier.

- Masking tape will make it easier to paint up to the edge of another surface.

- An edge roller will minimize masking and cut job time.

- An airless paint sprayer can dramatically cut application time in larger rooms though it adds to the prep and cleanup.

Potty Training

If you're doing much painting, invest in a lamb's wool roller, typically less than $15. Not only will it hold more paint and apply smoother, it also is easier to clean up than other rollers.

Preparing Surfaces for Paint

Once you've selected the paint and tools you need, the next job is to prepare the surfaces for paint. Fortunately, many of the surfaces are new and relatively clean. However, masonry foundation walls will need a thorough cleaning and preparation to accept new paint.

Even new bathroom surfaces will have dust and splatters from the work that's been going on. Use TSP (trisodium phosphate) or another cleaner recommended by your paint supplier to thoroughly clean the surface. Note that most cleaners will need a final rinse with clean water. Remem-ber to wear protective gloves when working with cleaners. As with any chemicals and cleaning agents, read and follow the manu-facturer's dir-ections.

If any patching is required to masonry, dry-wall, or other surfaces, now is the time to take care of it. Follow the directions on the masonry patch container, removing loose debris and apply-ing the patch in layers until you have a smooth,

dry surface. You can patch drywall with one of the popular patch kits and a container of wet patch.

Some paint manufacturers suggest a primer coat before the paint is applied. A primer seals the surface and gives the paint a good surface to which to bond. In addition, primer can help transition from the wall color to the chosen paint color. A good primer can minimize the number of paint coats required, thus saving time.

Painting a room typically means painting only specific surfaces—not window glass, door trim, flooring, and other surfaces. The best way to not get paint on an adjacent surface is to mask it, or use masking tape to cover the first inch or two of the surface that doesn't need painting. A 2-in. masking tape is most popular. If necessary, attach the tape to the edge of a plastic drop-cloth to keep paint from larger surfaces such as flooring.

Applying Paint

It's time to start painting. Here are some tips to make the job easier:

- Paint the ceiling before painting the walls.

- Use a trim or chisel brush to "cut in" or paint areas next to moldings and in corners where the roller can't easily get.

- Paint from top to bottom with smooth strokes.

- Once you start painting a wall don't stop until the wall is done so the paint dries evenly.

- Use a brush or roller extension when painting a ceiling, or use a *safe* platform that brings you closer to the surface.

- Use a paint shield, a flat metal or plastic piece with a handle, to shield adjacent surfaces from paint spatter.

- Don't dip the paint brush more than half way into the paint—let the bristles soak up paint, then wipe excess from the tip.

◆ Wipe excess paint from the roller before lifting it from the pan to minimize dripping and spatter.

◆ Roll paint on the wall in an M pattern about 3 ft. square, then add crosswise strokes to fill the pattern.

Most interior bathroom walls will be made of drywall (see Chapter 16). One reason why drywall is so popular is because of its paintability. The paper cover is smooth and readily absorbs paint. It frequently doesn't require a primer coat, saving time and expense. The downside of drywall is that the paint dries quickly. It's also an advantage, but it means that you can't stop in the middle of a wall for lunch. You should finish painting to a corner or, if possible, the entire room including trim. Otherwise, you may be able to see where you took a break. However, glossy paints tend to show touch-ups more than flat paints do.

Depending on your bathroom design you can paint the wood trim, doors, and windows the same color as the room or an accent color. In addition, you can decide to apply a wood stain, a paintlike product that has less pigment or color in it, allowing the wood's grain to show through. Some trim woods have little grain and should be painted, but oak or other wood with character looks best when stained.

Painting masonry, concrete, or concrete blocks is similar to painting any other flat surface. However, because masonry is porous, it typically requires a primer coat so the paint will adhere. Good quality masonry primer also can cover up any water stains. In addition, some masonry primers seal the surface against moisture. If painting masonry, select rollers specifically designed for the job. They can help work the primer or paint into the surface.

Cleaning Up

Cleaning up latex paint is relatively easy because it is water-soluble. You can soak hardened brushes until they are soft, then rinse them thoroughly to remove excess paint. Remember that brushes and rollers are designed to absorb lots of paint so rinsing the paint out may take awhile. Run water over and through tools until the water runs clear.

Potty Training

To eliminate cleanup after painting with latex, use disposable brushes and pads made from various foam products. They are inexpensive compared to professional painting brushes that you may only use once or twice a year.

Cleaning up after applying an oil-based paint is a little more work. It requires a paint solvent. Wear protective gloves and follow directions on the container.

Wallpapering

For instant decoration, apply wallpaper to bathroom walls along side or instead of paint. Wallpapering used to be a difficult job that required finesse and a few @#&*% words. Today's wall covering products are much easier to apply, especially the ones with adhesive already on the back side.

Potty Training

If you wallpaper, make sure the materials you select are appropriate for moist areas such as bathrooms. Otherwise colors may run and moisture will seep through the paper, requiring premature replacement.

The most difficult part of installing wallpaper is aligning it so that seams and any vertical patterns match up. You don't want to be able to see the seams.

The wall covering product you purchase will have installation instructions. However, most are installed in the same manner. The first sheet is installed vertically next to an inconspicuous corner such as behind the room's door. Mark the wall with a vertical starting line and align the paper's edge to it. Continue around the room, aligning and overlapping each sheet to match the pattern. As you go, carefully cut out around windows, doors, and plumbing and electrical fixtures.

Wallpaper installation tools.

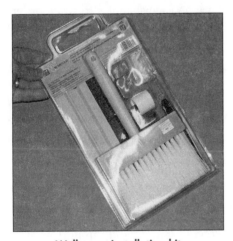

Wallpaper installation kit.

Preparing for wallpaper means making sure the wall surface is clean, dry, and smooth. Use TSP or a substitute to clean dirty surfaces. Otherwise, use a vacuum cleaner and a brush to remove any dust particles left over from remodeling your bathroom.

Waste Lines

Want to save a little time, effort, and money when wallpapering your bathroom? Install a prepasted border to add decorative color and style without all the work. Install the border at the top edge of the wall, making sure that the product you buy will stand up to a moist room.

Installing Trim

Trim is a necessity in that it covers the rough edges of construction. It is applied around doors and windows to cover the frame. For example, you made a rectangular hole in the wall, then inserted a prehung door in the rough opening. That's where trim can help. It also covers the edge between the walls and floor. Trim is multi-functional, too, offering a decorative accent.

Trim is also a luxury. Special trim such as *wainscoting* decorates walls. Ornate ceiling *molding* adds beauty to a room. Some prehung doors and windows include standard trim molding.

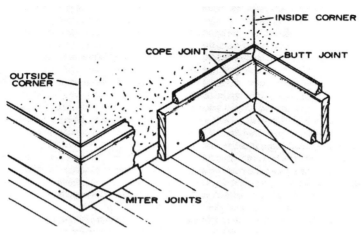

Common baseboard trim components.

Bathroom Words

Wainscoting is wood panels, boards, or other materials that cover the lower part of a wall, typically about 3 feet up from the floor. **Molding** is a wood or plastic strip that has a curved or projecting surface and is used for decorative purposes.

Because finishing your bathroom is more like new construction than remodeling, you can do what builders do: Paint, *then* trim. That saves you the task of masking off door, window, and floor trim as you paint walls.

Modern trim is relatively easy to install. That's because trim pieces come in standard designs and lengths to make construction easier. In many cases, all that's needed is to cut the trim to length and fasten it in place.

Only basic tools are needed to install molding:

◆ Finish hammer and nail set
◆ Miter box
◆ Miter, back, or coping saw
◆ Finish nails

Exact installation depends on the design of the molding. Simple door and window molding butts a vertical piece against a horizontal one. There are no special angles to cut. Corner molding typically requires an angle cut; that's where the miter box comes in.

A miter box is a tool that allows you to make perfect angle cuts. It's a fine-toothed saw and a three-sided box with slits that align the saw for 45-, 90-, and other popular angles. Wood and plastic miter boxes are adequate for simple angle cuts. If you plan to do more complex angles, opt for a steel miter box or even a power miter box. Here are some tips on cutting and installing trim molding:

◆ Practice cutting and fitting trim on scrap wood.
◆ Attach the miter box to a rigid surface.
◆ Make sure your saw is sharp and the sides are smooth so it tracks easily in the miter box notches.
◆ Start with the easiest trim and work up to the pieces that are more difficult.
◆ Apply the first coat of paint to cut trim and the second or touch-up coat once it is installed.

◆ Splice longer trim runs by cutting joint pieces at an angle.

Trim is cut either perpendicular (90 degrees) or diagonal (45 degrees) to the long edge.

Fasten trim with finish nails (the ones with a small head), use a nail set to indent them below the wood surface, then fill in the hole with wood filler and paint.

Measure the trim. If in doubt, cut long.

Once cut, place the trim and verify fit.

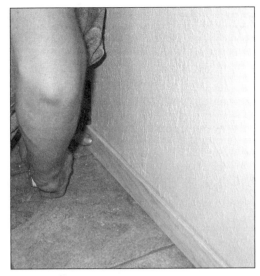

Nail the trim in place with finish nails.

Accessories

Bathroom accessories include towel bars, mirrors, and frames. Installation of these components is relatively easy.

Heads Up! For safety's sake, have someone help you install the mirror, especially if it is large. One person holds the mirror in place while the other measures and marks.

To install a wall-mounted mirror:

1. Place the mirror against the wall at the approximate location where you want to install it.
2. Use a measuring tape and a level to select the final location for the mirror.
3. Mark the corners of the mirror on the wall, then set the mirror aside.

4. Use an electronic stud-finder to locate wall studs into which the bottom mirror clips can be attached. If studs aren't conveniently located, install screw anchors into the drywall.

5. Install the bottom mirror clips.

6. Place the mirror on the bottom clips and adjust the side-to-side location of the mirror.

7. Install the top mirror clips directly above the bottom clips.

8. Install the side mirror clips.

9. Tighten all mirror clips for a snug fit.

Towel bars and picture frames are easier to install than mirrors. If possible, install bars and heavier frame hooks directly into wall studs. If there are no studs available at the location where you want to install the accessories, use anchors. Plastic anchor sleeves are used for lightweight accessories. Molly bolts and toggle bolts are better choices for towel bars and heavier frames and mirrors.

There are many other bathroom accessories that you may want to install. Most are either self-explanatory or come with instructions. For example, a towel bar kit typically includes wall plates, the bar, and mounting screws with instructions for installation. Most kits also include a plastic anchor sleeve for installing the unit into drywall. Good advice: Use toggle bolts and anchors for a firmer installation.

Enjoy your remodeled bathroom!

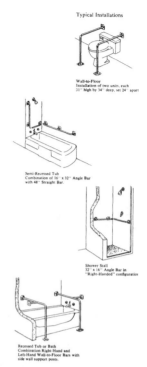

Typical Installations

Wall-to-Floor
Installation of two units, each
31" high by 34" deep, set 24" apart

Semi-Recessed Tub
Combination of 16" x 32" Angle Bar
with 48" Straight Bar.

Shower Stall
32" x 16" Angle Bar in
"Right-Handed" configuration

Recessed Tub or Bath
Combination Right-Hand and
Left-Hand Wall-to-Floor Bars with
side wall support posts.

Typical installation of ADA (Americans with Disabilities Act)–compliant and other assistive hardware for the bathroom.

The Least You Need to Know

- One of the easiest and most productive bathroom remodel jobs is painting walls and ceiling.
- Make sure you select moisture-resistant paint for bathroom surfaces.
- Install moisture-resistant wallpaper for a more decorative bathroom.
- Hanging a mirror and installing other accessories is made easier if you have a helper.

Bathroom Remodeling Glossary

ABS (acrylonitrile-butadine-styrene)   Material used for rigid black plastic pipe in DWV systems.

access hole   An opening in a ceiling or floor that provides access to the attic or crawl space.

adjustable-rate mortgage (ARM)   A mortgage in which the rates of interest and payment change periodically, based on a standard rate index. In most cases, the ARM has a cap or limit on how much the interest rate may increase.

air-dried lumber   Lumber that has been dried naturally by air with a minimum moisture content of 15 to 20 percent.

apron   The flat piece of inside trim that is placed against the wall directly under the sill of a window.

armored cable   A flexible metal-sheathed cable used for indoor wiring. Commonly called BX cable.

backhoe   A machine that digs deep, narrow trenches for foundations and drains.

balloon mortgage   A real estate loan in which some portion of the debt remains unpaid at the end of the term, resulting in a single, large payment due at the term's end.

baluster   A vertical member of the railing of a stairway, deck, balcony, or porch.

band joist   A joist nailed across the ends of the floor or ceiling joists. Also called a *rim joist*.

base molding   A strip of wood used to trim the upper edge of a baseboard.

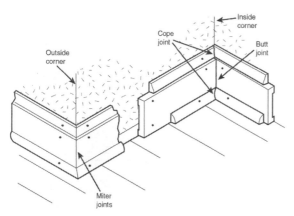

Cutting and installing wall base trim.

baseboard A trim board placed at the base of the wall, next to the floor.

bay window Any window—square, round, or polygonal—projecting outward from the wall of a structure.

beam A structural member, usually steel or heavy timber, used to support floor or ceiling joists or rafters.

bearing wall A wall that supports any vertical load in addition to its own weight.

blind-nailing Nailing in such a way that the nail heads are not visible on the face of the work.

blue lines or **blueprints** Reproductions of original construction documents that produce blue lines on a white background.

board foot A unit of lumber equal to a piece 1 foot square and 1 in. thick; 144 cubic in. of wood.

Nominal size (in.)	Actual length in feet								
	8	10	12	14	16	18	20	22	24
1 × 2		1²/3	2	2¹/3	2²/3	3	3¹/2	3²/3	4
1 × 3		2¹/2	3	3¹/2	4	4¹/2	5	5¹/2	6
1 × 4	2³/4	3¹/3	4	4²/3	5¹/3	6	6²/3	7¹/3	8
1 × 5		4¹/6	5	5⁵/6	6²/3	7¹/2	8¹/3	9¹/6	10
1 × 6	4	5	6	7	8	9	10	11	12
1 × 7		5⁵/8	7	8¹/6	9¹/3	10¹/2	11²/3	12⁵/6	14
1 × 8	5¹/3	6²/3	8	9¹/3	10²/3	12	13¹/3	14²/3	16
1 × 10	6²/3	8¹/3	10	11²/3	13¹/3	15	16²/3	18¹/3	20
1 × 12	8	10	12	14	16	18	20	22	24
1¹/4 × 4		4¹/6	5	5⁵/6	6²/3	7¹/2	8¹/3	9¹/6	10
1¹/4 × 6		6¹/4	7¹/2	8³/4	10	11¹/4	12¹/2	13³/4	15
1¹/4 × 8		8¹/3	10	11²/3	13¹/3	15	16²/3	18¹/3	20
1¹/4 × 10		10⁵/12	12¹/2	14⁷/12	16²/3	18³/4	20⁵/6	22¹¹/12	25
1¹/4 × 12		12¹/2	15	17¹/2	20	22¹/2	25	27¹/2	30
1¹/2 × 4	4	5	6	7	8	9	10	11	12
1¹/2 × 6	6	7¹/2	9	10¹/2	12	13¹/2	15	16¹/2	18
1¹/2 × 8	8	10	12	14	16	18	20	22	24
1¹/2 × 10	10	12¹/2	15	17¹/2	20	22¹/2	25	27¹/2	30
1¹/2 × 12	12	15	18	21	24	27	30	33	36
2 × 4	5¹/3	6²/3	8	9¹/3	10¹/3	12	13¹/3	14²/3	16
2 × 6	8	10	12	14	16	18	20	22	24
2 × 8	10²/3	13¹/3	16	18²/3	21¹/3	24	26²/3	29¹/3	32
2 × 10	13¹/3	16²/3	20	23¹/3	26²/3	30	33¹/3	36²/3	40
2 × 12	16	20	24	28	32	36	40	44	48
3 × 6	12	15	18	21	24	27	30	33	36
3 × 8	16	20	24	28	32	36	40	44	48
3 × 10	20	25	30	35	40	45	50	55	60
3 × 12	24	30	36	42	48	54	60	66	72
4 × 4	10²/3	13¹/3	16	18²/3	21¹/3	24	26²/3	29¹/3	32
4 × 6	16	20	24	28	32	36	40	44	48
4 × 8	21¹/3	26²/3	32	37¹/3	42²/3	48	53¹/3	58²/3	64
4 × 10	26²/3	33¹/3	40	46²/3	53¹/3	60	66²/3	73¹/3	80
4 × 12	32	40	48	56	64	72	80	88	96

Calculating board feet.

board lumber Yard lumber less than 2 in. thick and 2 or more in. wide.

bottom plate See *soleplate.*

box A metal or plastic container for electrical connections. More correctly called a *junction box.*

brace A piece of lumber or metal attached to the framing of a structure diagonally at an angle less than 90 degrees, providing stiffness or support.

branch In a plumbing or heating system, any part of the supply pipes connected to a fixture.

bridging Narrow wood or metal members placed on the diagonal between joists. Braces the joists and spreads the weight load.

builder's paper Usually, asphalt-impregnated paper of felt used in wall and roof construction; prevents the passage of air and moisture. Also known as building paper.

building code   The collection of legal requirements for the construction of buildings.

building drain   The lowest horizontal drainpipe in a structure. Carries all waste out to the sewer.

building permit   A permit issued by an appropriate government authority allowing construction of a project in accordance with approved specifications and drawings.

cable   An electricity conductor made up of two or more wires contained in an overall covering.

cap plate   The framing member nailed to the top plates of stud walls to connect and align them. The uppermost of the two top plates, sometimes called the double-top plate.

carriage   In a stairway, the supporting member to which the treads and risers are fastened. Also called a *stringer*.

casement   A window sash on hinges attached to the sides of a window frame. Such windows are called casement windows.

casing   Moldings of various widths and forms; used to trim door and window openings between the jambs and the walls.

caulk   Viscous material used to seal joints and make them water- and airtight.

caulk gun   A tool used to apply caulk.

cellar   A room or group of rooms below or predominantly below grade, usually under a building.

check valve   A valve that lets water flow in only one direction in a pipe system.

circuit   The path of electric current as it travels from the source to the appliance or fixture and back to the source.

circuit breaker   A safety device used to interrupt the flow of power when the electricity exceeds a predetermined amount. Unlike a fuse, you can reset a circuit breaker.

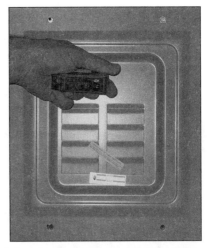

Circuit breakers, rated by the electrical current (in amps) they carry, are installed in a service panel.

cleanout   An easy-to-reach and easy-to-open place in a DWV (drain-waste-vent) system where obstructions can be removed or a snake inserted.

column   A vertical support—square, rectangular, or cylindrical—for a part of the structure above it.

common rafter   One of the parallel rafters, all the same length, that connect the eaves to the ridge board.

concrete   A mixture of aggregates and cement that hardens to a stonelike form and is used for foundations, paving, and many other construction purposes.

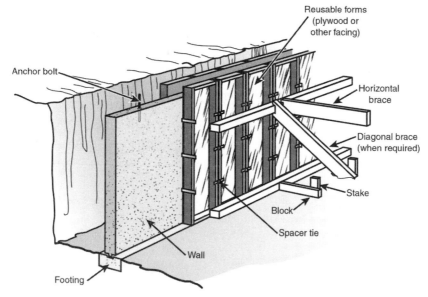

Anchor bolt

Reusable forms (plywood or other facing)

Horizontal brace

Diagonal brace (when required)

Stake

Block

Spacer tie

Wall

Footing

Construction of a concrete wall.

conductor   Any low-resistance material, such as copper wire, through which electricity flows easily.

conduit   A metal, fiber, or plastic pipe or tube used to enclose electric wires or cables.

conventional mortgage   An agreement between a buyer and a seller with no outside backing such as government insurance or guarantee.

cost per square foot (CPSF)   The figure obtained by dividing the total cost of construction or remodeling by the square feet in the area.

countersinking   Sinking or setting a nail or screw so that the head is flush with or below the surface.

cove molding   A molding with a concave face. Usually used to trim or finish interior corners.

CPVC (chlorinated polyvinyl chloride) The rigid white or pastel-colored plastic pipe used for supply lines.

crawlspace   A shallow space between the floor joists and the ground; usually enclosed by the foundation wall.

cripple studs   Short studs surrounding a window or between the top plate and end rafter, in a gable end or between the foundation and subfloor. Also called *jack studs.*

cross-bridging   Diagonal bracing between adjacent floor joists, placed near the center of the joist span to prevent joists from twisting.

crown molding   A convex molding used horizontally wherever an interior angle is to be covered (usually at the top of a wall, next to the ceiling).

current   The movement or flow of electrons, which provides electric power. The rate of electron flow as measured in amperes.

d   See *penny.*

dado joint   A joint where a dado or groove is cut in one piece of wood to accept the end of another piece.

dead load   The weight of the permanent parts of a structure that must be supported by the other parts of the structure. Does not include the weight of the people, furniture, and other things that occupy the building.

details Drawings used to clarify complicated construction features.

dimension lumber Yard lumber from 2 in. up to and including 4 in. thick and 2 or more in. wide. Includes joists, rafters, studs, plank, and small timbers.

direct-nailing Driving nails so that they are perpendicular to the surface or joint of two pieces of wood. Also called *face-nailing*.

doorjamb The case that surrounds a door. Consists of two upright side pieces called side jambs and a top horizontal piece called a head jamb.

draw request Monthly request by a contractor to be paid for the materials and labor installed into the project during the previous 30 days, to be drawn from the construction loan.

drywall Panels consisting of a layer of gypsum plaster covered on both sides with paper, used for finishing interior walls and ceilings. Also called *wallboard*, *gypsum wallboard*, and *Sheetrock*, a trade name.

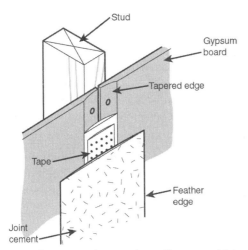

Drywall is seamed using drywall tape and joint cement.

ducts Pipes that carry air from a furnace or air conditioner to the living areas of a structure.

DWV (drain-waste-vent) An acronym referring to all or part of the plumbing system that carries waste water from fixtures to the sewer and gases to the roof.

edge grain Lumber that has been sawed parallel to the pith of the log and approximately at right angles to the growth rings.

elevations Representational drawings of interior and exterior walls to show finish features.

expansion joint A fiber strip used to separate blocks or units of concrete to prevent cracking due to expansion as a result of temperature changes. Often used on a larger concrete foundation and floor slabs.

face-nailing Driving nails so that they are perpendicular to the surface or joint of two pieces of wood. Also called *direct-nailing*.

finish carpentry The fine work—such as that for doors, stairways, and moldings—required to complete a building.

finish electrical work The installation of the visible parts of the electrical system, such as the fixtures, switches, plugs, and wall plates.

finish plumbing The installation of the attractive visible parts of a plumbing system such as plumbing fixtures and faucets.

fire-stop A solid, tight piece of wood or other material to prevent the spread of fire and smoke. In a frame wall, usually a piece of 2 × 4 cross-blocking between studs.

fitting In plumbing, any device that connects pipe to pipe or pipe to fixtures.

fixture In plumbing, any device that is permanently attached to the water system of a house. In electrical work, any lighting device attached to the surface, recessed into, or hanging from the ceiling or walls.

flange A connection between two plumbing components such as a toilet and a waste pipe.

floor plan A representational drawing of everything that constitutes the house.

footing The rectangular concrete base that supports a foundation wall or pier or a retaining wall. Usually wider than the structure it supports.

forms The temporary structure, usually of wood, that supports the shape of poured concrete until it is dry.

foundation The supporting portion of a structure below the first-floor construction or below grade, including the footings.

frame The enclosing woodwork around doors and windows. Also, the skeleton of a building; lies under the interior and exterior wall coverings and roofing.

furring strips Narrow strips of wood attached to walls or ceilings; forms a true surface on which to fasten other materials.

fuse A safety device for electrical circuits; interrupts the flow of current when it exceeds predetermined limits for a specific time period. Two popular types are cartridge and screw-in fuses.

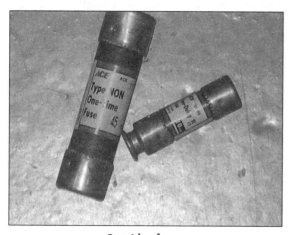

Cartridge fuses.

Screw-in fuses.

general conditions A listing of the requirements and understandings upon which a remodeling contract is based.

girder A large beam of steel or wood; supports parts of the structure above it.

grade The ground level surrounding a structure. The natural grade is the original level. The finished grade is the level after the structure is completed.

grain The direction, size, arrangement, appearance, or quality of fibers in wood.

ground Connected to the earth or something serving as the earth, such as a cold-water pipe. The ground wire in an electrical circuit is usually bare or has green insulation.

ground fault circuit interrupter (GFCI) An electrical safety device that senses any shock hazard and shuts off a circuit or receptacle.

grout Mortar that can flow into the cavities and joints of any masonry work, especially the filling between tiles and concrete blocks.

gypsum wallboard Panels consisting of a layer of gypsum plaster covered on both sides with paper, used for finishing interior walls and ceilings. Also called *wallboard*, *drywall*, and *Sheetrock*.

hanger   Any of several types of metal devices for supporting pipes, framing members, or other items. Usually referred to by the items they are designed to support—for example, joist hanger or pipe hanger.

hardboard   A synthetic wood panel made by chemically converting wood chips to basic fibers and then forming the panels under heat and pressure. Also called *Masonite*, a brand name.

hardwood   The wood of broadleaf trees, such as maple, oak, and birch. Although hardwood is usually harder than softwood, the term has no actual reference to the hardness of the wood.

header   A horizontal member over a door, window, or other opening; supports the members above it. Usually made of wood, stone, or metal. Also called a *lintel*. Also, in the framing of floor or ceiling openings, a beam used to support the ends of joists.

hot wire   In an electrical circuit, any wire that carries current from the power source to an electrical device. The hot wire is usually identified with black, blue, or red insulation, but it can be any color except white or green.

insulation   Any material that resists the conduction of heat, sound, or electricity.

insulation board   A structural building board made of coarse wood or cane fiber in ½-in. and ²⁵⁄₃₂-in. thicknesses. It can be obtained in various size sheets in various densities, and with several treatments.

interior finish   Any material (wall coverings and trim, for example) used to cover the framing members of the interior of a structure.

jack studs   See *cripple studs*.

jamb   The frame surrounding a door or window; consists of two vertical pieces called side jambs and a top horizontal piece called a head jamb.

joist   One of a series of parallel beams, usually 2 in. in thickness, used to support floor and ceiling loads and supported in turn by larger beams, girders, or bearing walls.

junction box   A metal or plastic container for electrical connections. Sometimes just called a *box*.

kiln-dried lumber   Lumber that has been kiln dried, often to a moisture content of 6 to 12 percent. Common varieties of softwood lumber, such as framing lumber, are dried to a somewhat higher moisture content.

laminate   To form a panel or sheet by bonding two or more layers of material. Also, a product formed by such a process—plastic laminate used for countertops, for example.

landing   The platform between flights of stairs or at the end of a stairway.

lath   A building material of metal, gypsum, wood, or other material; used as a base on which to apply plaster or stucco.

layout   Any drawing showing the arrangement of structural members or features. Also the act of transferring the arrangement to the site.

level   The position of a vertical line from any place on the surface of the earth to the center of the earth. Also, the horizontal position parallel to the surface of a body of still water. Also, a device used to determine when surfaces are level or plumb.

linear measure   Any measurement along a line.

live load   All loads on a building not created by the structure of the building itself; the furniture, people, and other things that occupy the building.

loan-to-value (LTV) ratio   The relationship between the amount of a mortgage loan and the value of the property.

lumber   A wood product manufactured by sawing, resawing, and passing wood lengthwise through a standard planing machine, and then crosscutting to length.

main drain   In plumbing, the pipe that collects the discharge from branch waste lines and carries it to the outer foundation wall, where it connects to the sewer line.

main vent In plumbing, the largest vent pipe to which branch vents may connect. Also called the *vent stack*.

Masonite A brand name for *wallboard*.

masonry Stone, brick, concrete, hollow tile, concrete block, gypsum, block, or other similar building units or materials bonded together with mortar to form a foundation, wall, pier, buttress, or similar mass.

mastic A viscous material used as an adhesive for setting tile or resilient flooring.

miter box A tool that guides a saw in making miter or angle cuts.

mortar A mixture of sand and Portland cement; used for bonding bricks, blocks, tiles, or stones.

mortgage An agreement between a lender and a buyer using real property as security for the loan.

mudsill The lowest member in the framing of a structure; usually 2-by lumber bolted to the foundation wall on which the floor joists rest. Also called a *sill plate*.

neutral wire In a circuit, any wire that is kept at zero voltage. The neutral wire completes the circuit from source to fixture or appliance to ground. The covering of neutral wires is always white.

nipple In plumbing, any short length of pipe externally threaded on both ends.

NM cable Nonmetallic sheathed electric cable used for indoor wiring. Also known by the brand name *Romex*.

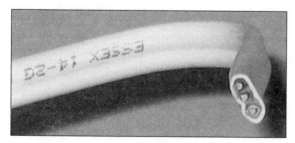

Nonmetallic (NM) cable includes two or more insulated wires and one bare ground wire.

nominal size The size designation of a piece of lumber before it is planed or surfaced. If the actual size of a piece of surfaced lumber is $1\frac{1}{2} \times 3\frac{1}{2}$ in., it is referred to by its nominal size: 2×4.

Nominal size (in.)	American standard (in.)
1 × 3	25/32 × 25/8
1 × 4	25/32 × 35/8
1 × 6	25/32 × 55/8
1 × 8	25/32 × 71/2
1 × 10	25/32 × 91/2
1 × 12	25/32 × 111/2
2 × 4	15/8 × 35/8
2 × 6	15/8 × 55/8
2 × 8	15/8 × 71/2
2 × 10	15/8 × 91/2
2 × 12	15/8 × 111/2
3 × 8	25/8 × 71/2
3 × 10	25/8 × 91/2
3 × 12	25/8 × 111/2
4 × 12	35/8 × 111/2
4 × 16	35/8 × 151/2
6 × 12	51/2 × 111/2
6 × 16	51/2 × 151/2
6 × 18	51/2 × 171/2
8 × 16	71/2 × 151/2
8 × 20	71/2 × 191/2
8 × 24	71/2 × 231/2

Nominal sizing of lumber.

nonbearing wall A wall supporting no load other than its own weight.

nosing The part of a stair tread that projects over the riser. Also, the rounded edge on any board.

on center Referring to the spacing of joists, studs, rafters, or other structural members as measured from the center of one to the center of the next. Usually written o.c. or OC.

oriented strand board (O.S.B.) A panel material of wood flakes compressed and bonded together with phenolic resin. Used for many of the same applications as plywood. Also known as structural flakeboard.

outlet In a wall, ceiling, or floor, a device into which the plugs on appliance and extension cords are placed to connect them to electric power. Properly called a *receptacle*.

panel A large, thin board or sheet of construction material. Also a thin piece of wood or plywood in a frame of thicker pieces, as in a panel door or wainscoting.

parquet A type of wood flooring in which small strips of wood are laid in squares of alternating grain direction. Parquet floors are now available in ready-to-lay blocks to be put down with mastic. Also any floor with an inlaid design of various woods.

particleboard A form of composite board or panel made of wood chips bonded with adhesive.

partition A wall that subdivides any room or space within a building.

penny As applied to nails, it originally indicated the price per hundred. The term now serves as a measure of nail length and is abbreviated by the letter *d*.

Common Wire Nails

Size	Length	Gauge	Approx. No. to Lb.	Size	Length	Gauge	Approx. No. to Lb.
2D	1 In.	No. 15	876	10D	3 In.	No. 9	69
3D	1 1/4	14	568	12D	3 1/4	9	63
4D	1 1/2	12 1/2	316	16D	3 1/2	8	49
5D	1 3/4	12 1/2	271	20D	4	6	31
6D	2	11 1/2	181	30D	4 1/2	5	24
7D	2 1/4	11 1/2	161	40D	5	4	18
8D	2 1/2	10 1/4	106	50D	5 1/2	3	14
9D	2 3/4	10 1/4	96	60D	6	2	11

Flooring Brads

Size	Length	Gauge	Approx. No. to Lb.
6D	2 In.	No. 11	157
7D	2 1/4	11	139
8D	2 1/2	10	99
9D	2 3/4	10	90
10D	3	9	69
12D	3 1/4	8	54
16D	3 1/2	7	43
20D	4	6	31

Finishing Nails

Size	Length	Gauge	Approx. No. to Lb.
2D	1 In.	No. 16 1/2	1351
3D	1 1/4	15 1/2	807
4D	1 1/2	15	584
5D	1 3/4	15	500
6D	2	13	309
7D	2 1/4	13	238
8D	2 1/2	12 1/2	189
9D	2 3/4	12 1/2	172
10D	3	11 1/2	121
12D	3 1/4	11 1/2	113
16D	3 1/2	11	90
20D	4	10	62

Smooth & Barbed Box Nails

Size	Length	Gauge	Approx. No. to Lb.
2D	1 In.	No. 15 1/2	1010
3D	1 1/4	14 1/2	635
4D	1 1/2	14	473
5D	1 3/4	14	406
6D	2	12 1/2	236
7D	2 1/4	12 1/2	210
8D	2 1/2	11 1/2	145
9D	2 3/4	11 1/2	132
10D	3	10 1/2	94
12D	3 1/4	10 1/2	88
16D	3 1/2	10	71
20D	4	9	52
30D	4 1/2	9	46
40D	5	8	35

Casing Nails

Size	Length	Gauge	Approx. No. to Lb.
2D	1 In.	No. 15 1/2	1010
3D	1 1/4	14 1/2	635
4D	1 1/2	14	473
5D	1 3/4	14	406
6D	2	12 1/2	236
7D	2 1/4	12 1/2	210
8D	2 1/2	11 1/2	145
9D	2 3/4	11 1/2	132
10D	3	10 1/2	94
12D	3 1/4	10 1/2	87
16D	3 1/2	10	71
20D	4	9	52
30D	4 1/2	9	46

Nail types and sizes.

Phillips head A kind of screw and screwdriver on which the diving mechanism is an X rather than a slot.

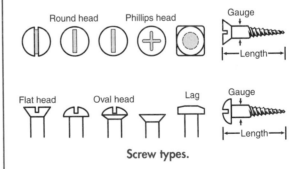

Screw types.

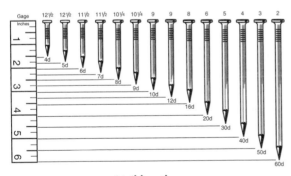

Nail lengths.

pier A column of masonry, usually rectangular, used to support other structural members. Often used as a support under decks.

pigtail A short length of electrical wire or group of wires.

plan The representation of any horizontal section of a structure, part of a structure, or the site of a structure; shows the arrangement of the parts in relation and scale to the whole.

plaster A mixture of lime, sand, and water plus cement for exterior cement plaster, and plaster of paris for interior smooth plaster used to cover the surfaces of a structure.

plasterboard See *wallboard*.

plate A horizontal framing member, usually at the bottom or top of a wall or other part of a structure, on which other members rest. The *mudsill*, *soleplate*, and *top plate* are examples.

plumb Exactly perpendicular; vertical.

plywood A wood product made up of layers of wood veneer bonded together with adhesive. It is usually made up of an odd number of plies set at a right angle to each other.

post A vertical support member, usually made up of only one piece of lumber or a metal pipe or I-beam.

principal and interest The monthly cost of a mortgage; principal is the amount borrowed—the difference between the cost of the home and the down payment—while interest is the charge made by the lender for lending the money. Abbreviated *PI*.

putty A soft, pliable material used for sealing the edges of glass in a sash or to fill small holes or cracks in wood.

PVC (polyvinyl chloride) A rigid, white, plastic pipe used in plumbing for supply and DWV systems.

quarter-round A convex molding shaped like a quarter circle when viewed in cross section.

radiant heating Electrically heated panels or hot-water pipes in the floor or ceiling that radiate heat to warm the room's surfaces.

receptacle In a wall, ceiling, or floor, an electric device into which the plugs on appliance and extension cords are placed to connect them to electric power. Also called an *outlet*.

register In a wall, floor, or ceiling, the device through which air from the furnace or air conditioner enters a room. Also any device for controlling the flow of heated or cooled air through an opening.

reinforcing bar Steel bars or wire mesh placed in concrete to increase its strength. Also called *rebar*.

retainage The amount (usually 10 percent) held back by an owner out of each payment to the general contractor, to be held as security that the work will be finished and to be paid when the work is complete.

rim joist See *band joist*.

ripping Sawing wood in the direction of the grain.

riser Each of the vertical boards between the treads of a stairway.

Romex A brand name for nonmetallic sheathed electric cable used for indoor wiring. Also called *NM cable*.

rough-in To install the basic, hidden parts of a plumbing, electrical, or other system while the structure is in the framing stage. Contrasts with installation of finish electrical work or plumbing, which consists of the visible parts of the system.

run In stairways, the front-to-back width of a single stair or the horizontal measurement from the bottom riser to the back of the top tread.

scale The proportion between two sets of dimensions. On building plans, the house is drawn smaller than the actual house, but in scale so that the proportions are the same. For example, when the scale is expressed as $\frac{1}{4}'' = 1'0''$, $\frac{1}{4}$ in. on the drawing equals 1 foot on the actual house.

section A drawing of part of a building as it would appear if cut through by a vertical plane.

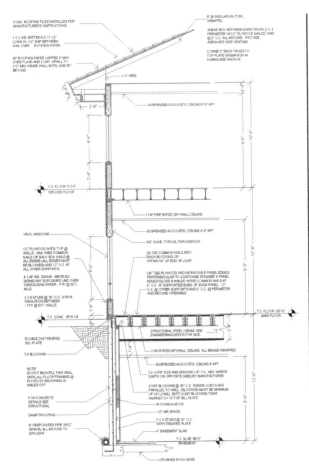

Typical construction section.

service panel The box or panel from which the electricity is distributed to the house circuits. It contains the circuit breakers and, usually, the main disconnect switch.

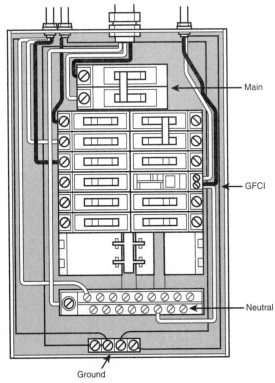

The service panel distributes electricity through circuits, each with its own breaker.

Sheetrock A commercial name for *wallboard*.

shim A thin wedge of wood, often part of a shingle, used to bring parts of a structure into alignment.

shoe molding A strip of wood used to trim the bottom edge of a baseboard.

shutoff valve In plumbing, a fitting to shut off the water supply to a single fixture or branch of pipe.

sill plate The lowest member in the framing of a structure; usually a 2-by board bolted to the foundation wall on which the floor joists rest. Also called a *mudsill*.

site plan Drawing of all the existing conditions on the lot, usually including slope and other topography, existing utilities, and setbacks. These drawings may be provided by the municipality.

slab   A concrete foundation or floor poured directly on the ground.

sleepers   Boards embedded in or attached to a concrete floor; serve to support and provide a nailing surface for a subfloor or finish flooring.

soffit   The underside of a stairway, cornice, archway, or similar member of a structure. Usually a small area relative to a ceiling.

soft costs   All the costs associated with the beginning of a remodeling project that purchase intangible items that cannot be resold, such as legal fees, architect and engineering fees, and loan fees.

soil stack   In the DWV system, the main vertical pipe. Usually extends from the basement to a point above the roof.

solderless connector   A product that establishes connection between two or more electrical conductors without solder. Also called a *wire nut*.

soleplate   In a stud wall, the bottom member, which is nailed to the subfloor. Also called a *bottom plate*.

solid bridging   A solid member placed between adjacent floor joists near the center of the span to prevent joists from twisting.

span   The distance between structural supports, such as walls, columns, piers, beams, girders, and trusses.

specifications   Written lists, instructions and general information that relate to the construction and make up a part of the total legal contract.

splash block   A small masonry block laid with the top close to the ground surface to receive roof drainage from downspouts and to carry it away from the foundation.

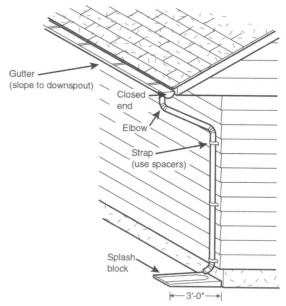

Gutter and downspout system including splash block.

square   A term used to describe an angle of exactly 90 degrees. Also a device to measure such an angle. Also a unit of measure equaling 100 square feet.

stringer   In a stairway, the supporting member to which the treads and risers are fastened. Also called a *carriage*.

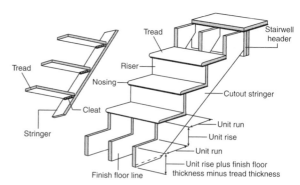

Components of a stairway, including the stringer.

strip flooring Wood flooring consisting of narrow, matched strips.

stucco A plaster of sand, Portland cement, and lime used to cover the exterior of buildings.

stud One of a series of wood or metal vertical framing members that are the main units of walls and partitions.

stud wall The main framing units for walls and partitions in a building, composed of studs; top plates; bottom plates; and the framing of windows, doors, and corner posts.

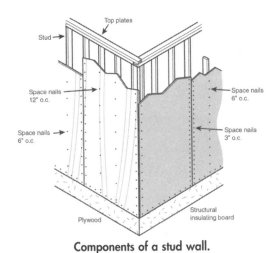

Components of a stud wall.

subfloor Plywood or oriented strand boards attached to the joists. The finish floor is laid over the subfloor. The subfloor also can be made of concrete.

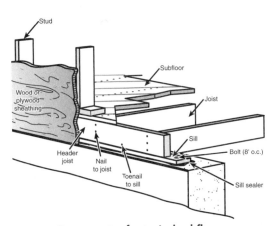

Components of a typical subfloor.

substantial completion The time at which the contractor feels that he or she has essentially finished the project, but before the final inspection.

suspended ceiling A system for installing ceiling tile by hanging a metal framework from the ceiling joists.

switch In electrical systems, a device for turning the flow of electricity on and off in a circuit or diverting the current from one circuit to another.

termite shield Galvanized steel or aluminum sheets placed between the foundation, pipes, or fences and the wood structure of a building; prevents the entry of termites.

threshold A shaped piece of wood or metal, usually beveled on both edges, that is placed on the finish floor between the side jamb; forms the bottom of an exterior doorway.

timber Pieces of lumber with a cross section greater than 4 × 6 in. Usually used as beams, girders, posts, and columns.

toenailing Driving a nail at a slant to the initial surface in order to permit it to penetrate into a second member.

tongue and groove A way of milling lumber so that it fits together tightly and forms an extremely strong floor or deck. Also, boards milled for tongue-and-groove flooring or decking that have one or more tongues on one edge and a matching groove or grooves on the other.

top plate In a stud wall, the top horizontal member to which the cap plate is nailed when the stud walls are connected and aligned.

trap In plumbing, a U-shaped drain fitting that remains full of water to prevent the entry of air and sewer gas into the building.

tread In a stairway, the horizontal surface on which a person steps.

trim   Any finish materials in a structure that are placed to provide decoration or to cover the joints between surfaces or contrasting materials. Door and window casings, baseboards, picture moldings, and cornices are examples of trim.

underlayment   The material placed under the finish coverings of floors to provide waterproofing as well as a smooth, even surface on which to apply finish material.

vapor barrier   Any material used to prevent the penetration of water vapor into walls or other enclosed parts of a building. Polyethylene sheets, aluminum foil, and building paper are the materials used most.

veneer   A thin layer of wood, usually one that has beauty or value, that is applied for economy or appearance on top of an inferior surface.

vent   Any opening, usually covered with screen or louvers, made to allow the circulation of air, usually into an attic or crawlspace. In plumbing, a pipe in the DWV system for the purpose of bringing air into the system.

vent stack   In plumbing, the largest vent pipe to which branch vents may connect. Also called the *main vent*.

wall plate   A decorative covering for a switch, receptacle, or other device.

wallboard   Panels consisting of a layer of gypsum plaster covered on both sides with paper, used for finishing interior walls and ceilings. Also called *gypsum wallboard*, *drywall*, and *Sheetrock*.

water-repellent preservative   A liquid designed to penetrate wood and impart water repellency and a moderate preservative protection.

weather stripping   Narrow strips of metal, fiber, plastic foam, or other materials placed around doors and windows; prevents the entry of air, moisture, or dust.

wire nut   A device that uses mechanical pressure rather than solder to establish a connection between two or more electrical conductors. Also called a *solderless connector*.

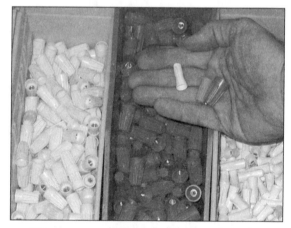

Wire nuts are available in a variety of sizes to match wire gauges.

Bathroom Remodeling Resources

This book is written to answer your primary questions about how to remodel your bathroom, offering specific options and processes. For additional information on designing, planning, financing, building, and enjoying your newly remodeled bathroom, refer to the more than 100 resources in this appendix.

The Internet is the best source of current information. It is also widely available in homes, businesses, libraries, and even coffee shops. Therefore, most resources in this appendix refer you to Internet websites that will include up-to-date general and specific information on a wide variety of building topics. Remember to add the prefix "www." to all URLs (uniform resource locators) in this appendix.

Valuable Online Resources

101HomeResources.com:
Comprehensive home decorating and building resources directory

BobVila.com:
Online resource for home remodeling and construction

Build.com:
Resource for builders, contractors, remodelers, and do-it-yourselfers

ChiefArchitect.com:
Professional drafting and design software by Advanced Relational Technology (ART), Inc.

DoItYourself.com:
Wealth of information for all do-it-yourselfers

FixItClub.com:
Fix-It Guides and resources for fixing household things

Freeware.com and Shareware.com:
Free and low-cost software programs

Google.com:
Search engine for finding things on the Internet

Homeplanner.com:
Tools for home design and remodeling

InsideSpaces.com:
For novice home remodelers

MulliganBooks.com:
Offering do-it-yourself remodeling books by Dan Ramsey and others

PunchSoftware.com:
Professional home design software by Punch Software

RecRoomFurniture.com:
Wide variety of furniture for your recreation room

StartRemodeling.com:
Extensive site for everything involving remodeling

YourCompleteHome.com:
Home plans and contractor directory

Remodeling and Construction Magazines

These and similar magazines are available through larger newsstands and online at the following URLs:

Better Homes & Gardens: bhg.com

Builder: builderonline.com

Canadian House & Home: canadianhouseandhome.com

Canadian Living: canadianliving.com

Coastal Living: coastallivingmag.com

Concrete Homes: concretehomesmagazine.com

Country Living: countryliving.com

Environmental Building News: ebuild.com

Fine Homebuilding: finehomebuilding.com

Good Housekeeping: goodhousekeeping.com

Home and Design: homeanddesign.com

House Beautiful: housebeautiful.com

Log Home Living: homebuyerpubs.com

Metal Construction News: moderntrade.com

Natural Home: naturalhomemagazine.com

Permanent Buildings & Foundations: pbf.org

Residential Architect: residentialarchitect.com

Southern Living: southern-living.com

Sunset: sunsetmagazine.com

This Old House: thisoldhouse.com

Traditional Building: traditionalbuilding.com

Wood Design & Building: wood.ca

Books and Videos

Baths: Your Guide to Planning and Remodeling. Des Moines, IA: Better Homes and Gardens Books, 1996.

Ching, Frank. *Building Construction Illustrated.* Hoboken, NJ: John Wiley, 2000.

Germer, Jerry. *Creating Beautiful Bathrooms.* Upper Saddle River, NJ: Creative Homeowner, 2001.

Kardon, Redwood, Michael Casey, and Douglas Hansen. *Code Check: A Field Guide to Building a Safe House.* Newtown, CT: Taunton Press, 2000.

National Electrical Code. Quincy, MA: National Fire Protection Association, 2001.

National Standard Plumbing Code. Falls Church, VA: National Association of Plumbing-Heating-Cooling Contractors, 2000.

Ogershok, Dave. *National Construction Estimator.* Los Angeles: Craftsman Book Co., 2002.

Ortho's All About Bathrooms. Des Moines, IA: Meredith Books, 1998.

Ramsey, Dan. *The Complete Idiot's Guide to Building Your Own Home.* Indianapolis, IN: Alpha Books, 2002.

———. *The Complete Idiot's Guide to Finishing Your Basement.* Indianapolis, IN: Alpha Books, 2003.

Ramsey, Dan, and Judy Ramsey. *If It's Broke, Fix It!* Indianapolis, IN: Alpha Books, 2003.

The Complete Guide to Bathrooms. Chanhassen, MN: Creative Publishing, 2003.

Vila, Bob and Howard, Hugh. *Bob Vila's Complete Guide to Remodeling Your Home.* New York, NY: Quill, 1999.

Books listed here are available through local bookstores or from MulliganBooks.com.

Remodeling Trade Association Resources Online

Air Conditioning Contractors of America: acca.org

American Concrete Institute: aci-int.org

American Hardware Manufacturers Association: ahma.org

American Institute of Architects: aiaonline.com

American Institute of Building Design: aibd.org

American Lighting Association: americanlightingassoc.com

American National Standards Institute: ansi.org

American Society for Testing and Materials: astm.org

American Society of Heating, Refrigerating & Air Conditioning Engineers: ashrae.org

American Society of Interior Designers: asid.org

Americans with Disabilities Act: usdoj.gov/crt/ada/adahom1.htm

Architectural Woodwork Institute: awinet.org

Associated Builders and Contractors: abc.org

Associated General Contractors of America: agc.org

Associated Soil and Foundation Engineers: asfe.org

Association of Home Appliance Manufacturers: aham.org

Brick Institute of America: bia.org

Building Officials and Code Administrators International: bocai.org

California Redwood Association: calredwood.org

Canadian Homebuilders Association: chba.ca

Canadian Institute of Plumbing and Heating: ciph.com

Canadian Standards Association: csa.ca

Canadian Window and Door Manufacturers Association: windoorweb.com

Cast Iron Soil Pipe Institute: cispi.org

Composite Panel Association and Composite Wood Council: pbmdf.com

Concrete Reinforcing Steel Institute: crsi.org

Construction Specifications Institute: csinet.org

Contractor License Reference Site: contractors-license.org

The Council of American Building Officials: cabo.org

Energy Efficient Building Association: eeba.org

The Engineered Wood Association: apawood.org

The Gypsum Association: gypsum.org

Hardwood Council: hardwoodcouncil.com

Hardwood Plywood and Veneer Association: erols.com/hpva

Home Builders Institute: hbi.org

Institute of Electrical and Electronics Engineers: ieee.org

Insulating Concrete Form Association: forms.org

International Conference of Building Officials: icbo.org

International Organization for Standardization: iso.ch

National Aggregates Association: nationalaggregates.org

National Air Duct Cleaners Association: nadca.com

National Association of Home Builders: nahb.com

National Association of the Remodeling Industry: nari.org

National Association of Women in Construction: nawic.org

National Concrete Masonry Association: ncma.org

National Electrical Contractors Association: necanet.org

National Electrical Manufacturers Association: nema.org

National Fire Protection Association: nfpa.org

National Hardwood Lumber Association: natlhardwood.org

National Institute of Building Sciences: nibs.org

National Institute of Standards and Technology: nist.gov

National Oak Flooring Manufacturers Association: nofma.org

National Pest Control Association: pestworld.org

National Rural Water Association: nrwa.org

National Stone Association: aggregates.org

National Tile Contractors Association: tile-assn.com

National Wood Flooring Association: woodfloors.org

National Wood Window and Door Association: nwwda.org

North American Insulation Manufacturers Association: naima.org

North American Steel Framing Alliance: steelframingalliance.com

Occupational Safety and Health Administration: osha.gov

Plumbing-Heating-Cooling Contractors National Association: phccweb.org

Portland Cement Association: concretehomes.com

Remodeling Association: remodelingassociation.com

Sheet Metal and Air Conditioning Contractors National Association: smacna.org

Southern Building Code Congress International: sbcci.org

Southern Forest Products Association: sfpa.org

Southern Pine Council: southernpine.com

Steel Door Institute: wherryassoc.com/ steeldoor.org

Structural Board Association: sba-osb.com

Underwriters Laboratories Inc.: ul.com

Western Red Cedar Lumber Association: cofi.org/WRCLA

Western Wood Products Association: wwpa.org

Wood Floor Covering Association: wfca.org

Primary Trade Association Addresses

Air Conditioning Contractors of America
1513 16th Street NW
Washington, DC 20036
202-483-9370

American Subcontractors Association
1004 Duke Street
Alexandria, VA 22314
703-684-3450

Associated Builders and Contractors, Inc.
729 15th Street NW
Washington, DC 20006
202-393-2040

Associated Specialty Contractors
7315 Wisconsin Avenue
Bethesda, MD 20814
301-657-3110

Builders Alliance
P.O. Box 20308
Seattle, WA 98102
206-323-1966

Engineering Contractors Association
8310 Florence Avenue
Downey, CA 90240
213-861-0929

Floor Covering Installation Contractors
Association
P.O. Box 2048
Dalton, GA 30722
404-226-5488

General Building Contractors Association
36 South 18th Street
Philadelphia, PA 19103
215-568-7015

Independent Electrical Contractors of America
P.O. Box 10379
Alexandria, VA 22310
703-549-7351

Independent Professional Painting Contractors
Association
P.O. Box 1759
Huntington, NY 11743
516-423-3654

Insulation Contractors Association of America
15819 Crabbs Branch Way
Rockville, MD 20855
301-590-0030

International Remodeling Contractors
Association
P.O. Box 17063
West Hartford, CT 06117
203-233-7442

Mason Contractors Association of America
33 S. Roselle Road
Schaumburg, IL 60193
847-301-0001

Mechanical Contractors Association of America
1385 Piccard Drive
Rockville, MD 20832
301-869-5800

Metal Construction Association
1101 14th Street NW
Washington, DC 20005
202-371-1243

National Association of Home Builders of the
United States
15th and M Street NW
Washington, DC 20005
202-822-0200

National Association of Minority Contractors
806 15th Street NW
Washington, DC 20005
202-347-8259

National Association of Plumbing-Heating-
Cooling Contractors
P.O. Box 6808
Falls Church, VA 22046
703-237-8100

National Association of Reinforcing Steel
Contractors
P.O. Box 280
Fairfax, VA 22030
703-591-1870

National Constructors Association
1730 M Street NW
Washington, DC 20036
202-466-8880

National Electrical Contractors Association
7315 Wisconsin Avenue NW
Washington, DC 20014
202-657-3110

National Insulation Contractors Association
99 Canal Center Plaza
Alexandria, VA 22314
703-683-6422

National Tile Contractors Association
P.O. Box 13629
Jackson, MS 39236
601-939-2071

Painting and Decorating Contractors of
America
3913 Old Lee Highway
Fairfax, VA 22030
703-359-0826

Poured Concrete Wall Contractors Association
825 E. 64th Street
Indianapolis, IN 46220
317-253-5655

Professional Construction Estimators
Association
P.O. Box 11626
Charlotte, NC 28220
704-522-6376

Sheet Metal and Air Conditioning Contractors
National Association
4201 Lafayette Center Drive
Chantilly, VA 22021
703-803-2980

Government Resources

Department of Housing and Urban
Development: hud.gov

Environmental Protection Agency: epa.gov

Federal Home Loan Mortgage Corporation:
freddiemac.com

Federal Housing Administration: hud.gov/fha/

Federal Housing Finance Board: fhfb.gov

Federal National Mortgage Association:
fanniemae.com

Small Business Administration: sba.gov

Veterans Administration Home Loan
Guarantee Program: homeloans.va.gov

Index

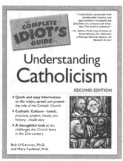

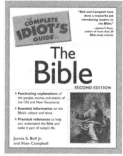